A Common Destiny

*To Héléna and Gaspard.*

# A Common Destiny

A Photographic Journey Through a Changing World

Cédric Delsaux

Introduction by Bill McKibben

The Monacelli Press

Foreword **Cédric Delsaux**

# "I wanted to undertake this voyage to the heart of our contradictions, to visit the hidden face of our good consciences."

**What are we going to do now,** here, in this succession of improbable places? We are feeble, caught in the snares of a megamachine. Passengers en route to a compromised destination.

I remember, as a child, having dreamed of distant worlds, of science fiction universes. But what else was I waiting for? Spaceships? Three-headed Martians? They were already here, present in their most perfect banality, in a most incredible reality. All the same, I remain fascinated by the universe that we have created, by the substances it is made of, its colors and shapes; it takes a portion of genius to invent it . . . and a portion of folly to live here.

At this point, in most large cities, we can go through life without ever putting a foot on earth: we touch only parking lots, corridors, offices, shopping centers. We no longer live on this planet—we have constructed another in place of the first. The most astonishing thing is that this hallucinatory world now seems real. The underlying ideology works perfectly well. In trying to make these artificial landscapes appear "natural," like a self-evident necessity, they have entered into our collective unconscious.

The problem, basically, is that this technological world, this machine world, this world of mass consumption, no longer needs us: the individual is a negligible residue. This world of numbers, quantifiable in units, in mathematical theories, in return on investment, is an objectifying world, without any sort of counterbalance to allow for even the possibility of "otherness." This world leads only to the filling of shelves and the emptying of spirits. All that remains is for each of us to save himself, to attempt to escape the steamroller, even if it means becoming a steamroller, and to parade around in our self-satisfaction.

We are so preoccupied with being recognized, with our own success, that we would rather risk the annihilation of our universe than personal failure. We want nothing which might spoil our comfort.

While we call for clear thinking on the part of others, we remain blind to our own contradictions. This society is so complex that we are all caught up in a Gordian knot of causes and consequences from which we can no longer emerge intact. To remain free becomes a chaotic journey. The compromised often either compromise themselves or turn to the margins of society; the flight ends inescapably in a dead end.

But we can't continue to hide for long: our world is no longer viable.

A single planet will not be sufficient to satisfy our ambitions. Our eyes are bigger than the Earth. We must change the ways we live, our dreams, our ideals. And unfortunately, if we undertake nothing, we might as well prepare for the worst.

The proposition of Olivier Bourgeois, producer and codirector of the documentary *Nous Resterons sur Terre/Here to Stay*, to accompany his crew as they filmed many of the same places featured here allowed me to begin verifying this hypothesis. Coincidentally, he was looking for a photographer whose work explored the same themes as the film. He gave me carte blanche to explore my perception of the world in my own way and without any pressure.

For a long time, I had wanted to undertake this voyage to the heart of our contradictions, to visit the hidden face of our good consciences. Finally, through the intervention of the film, I could set out to do it.

At first I had no plan, no certitude. I had only the constant sensation that "the world is sad and beautiful" as Giacomo Leopardi wrote, and that we had to continue to live here, whatever the cost. So I began creating an inventory of places symptomatic of our era. A partial and scattered inventory–from the start I rejected the notion of an exhaustive catalog.

The idea was to become inspired by the form of the documentary style but to distance myself from its concept of "objectivity."

On the contrary, it seemed to me that the project's appeal could come from its capacity to make us feel "unglued" from reality. I needed, then, to explore the symbolic potential of these places and to not limit myself to creating a collection of unrelated scenes.

It didn't matter to me what these factories made, what was constructed or abandoned there, because these places became the actors, the representatives of a collective history: that of the westernization of the world. From this angle, they assume the role of symbol or, at least, of witness. Witness to our capacity to abnegate human nature as well as witness to our pretension and recklessness. But I wanted more than anything to infuse these places equally with shadow, with nothingness, and with folly to make them more mysterious, more desirable. To give them the ability to bewitch us.

Stripped of their normal functions in this way, they became "floating signifiers" that were able to wander in my imagination until they formed a chain of heterogeneous forms, a sort of mythology of postmodern spaces.

Both real and imaginary, they opened a fissure in the core of this reality, prompting a shift in its landscape from which was thrust a form of poetry and of melancholy.

I could not keep from finding beauty there, at the very heart of the ugliness. There, I recognized the intimacy of my bond with this world that I love and loathe simultaneously. This tenuous, uncertain link that keeps me alive. This steady gaze that does not want to be dimmed. And the undeniable fragility that flows from it.

Fragility common to all who, like me, are reconciled to living here without answers, without God, and without weapons.

Introduction **Bill McKibben**

# "The set of desires we have constructed have done more than damage the planet. They have also damaged us."

There are no people here. Maybe six or seven scattered throughout these extraordinary images. A tiny figure in a raincoat beneath a waterfall. An out-of-focus worker in some factory. A small cluster of onlookers in a tour boat, busy photographing the icebergs melting around them.

And yet, of course, people are what the photographs are all about. Save for the first few images, which serve as a kind of baseline—river! mountain! ocean!–each pulsates with the waves of human desire, at a frequency not quite visible to the naked eye except in its effects on the world. Some of the desires are clearly innocent; for instance, that of an Amazonian tribe for some connection to the larger world (the outcome of which one can guess at with a shudder). Some, such as the desire to go "skiing" beneath an air-conditioner, are pathetic. Some of the desires are the result of manipulation, but no less deep because of it. Most of the desires are brute, almost inchoate—for food, for energy, for goods on a scale so enormous as to seem inhuman. But the desires are never inhuman—each tire in a vast dump is a testament to the thousands of trips it undertook, to the store or the office or the assignation. Each dipping nod of the oil rig driven by the hope for a burst of some particular pleasure, whether speed, light, or warmth.

Our industrial civilization has reached a scale, however, where our desires have become the central issue. Because of our sheer numbers, the size of each appetite is amplified: the trillions of miles covered by car, the billions of airline trips, the size of our houses. It is in our lifetimes—two or three decades out of the full span of human civilization—that those desires have reached a critical volume, large enough to erode all that had seemed safe and solid. And the numbers make the point.

Before the Industrial Revolution, the atmospheric concentration of carbon dioxide was 275 parts per million. That figure had scarcely wavered over the 10,000 years of human civilization—what it was for Shakespeare it was for the Buddha. It defined the world humans knew: a world, for instance, warm enough that the continental interiors were free of ice but cold enough at high elevations for glaciers to remain and provide a steady stream of summer water to slake Asia and South America and California.

When we began to burn coal and gas and oil in large quantities (and all modern desire requires coal and gas and oil), that atmospheric concentration began to rise. We never knew where the danger line was—always in the future, we hoped. But more $CO_2$ piled up in the atmosphere each year. At the moment, the number is about 390 parts per million and rising about 2 parts per million annually. That is the main difference between our time, when we have become flesh-colored machines for the combustion of fossil fuels, and those of Moses, Confucius, or Itzcóatl.

In the summer of 2007, sea ice in the Arctic began, suddenly and radically, to melt. By the season's end, there was 25 percent less ice than ever before in human history. This signaled that the line had been crossed, that our alteration of the earth was no longer a future problem but a present emergency—and very close to a fait accompli. In the months afterward, NASA's James Hansen, a renowned climatologist, and a team of fellow researchers cast this insight in numerical terms. Any value for carbon in the atmosphere greater than 350 parts per million, they said, was incompatible with a "planet similar to that on which civilization developed and to which life on Earth is adapted."

The photographs in this book are the pictures of a world above 350 parts per million, a world so out of scale that we are literally—and quickly—making life all but impossible for all of the planet's inhabitants. (The image of the hippos headed for the exit in the zoo takes on a surprising poignancy in this context.) Climate change has an almost existential quality. Invisible gases cause change just slow enough to escape the notice of the casual observer. You can see the evidence of it in certain images—in the form of, say, the black clouds over the cities of China. But mostly you can't see it; most of the time you have to infer it, and the brooding, heavy, menacing images presented in this book are pregnant with violent possibility.

When that violence erupts—when great storms power in off the oceans, when mosquitoes spread an outbreak of dengue in some Asian city, when rains flood Africa, when droughts parch whole continents—we are too busy dealing with the effects to think of causes. And when crises come fast enough, we give up and leave. The photographs of Chernobyl could as well be of New Orleans—or, tomorrow, of Bangladesh, of Somalia, of California. People have already abandoned towns in coastal Alaska, and whole island nations are beginning to pack up and move as sea level rises. Shortly after taking office in 2008, the president of the Maldives said, "If we are unable to save countries like [ours] it may be too late to save the rest of the world from the apocalyptic effects of self-reinforcing, runaway global warming." How confident was he that the world would change in time? He announced that his country was salting away a large share of each year's tourism proceeds in order to purchase a new homeland somewhere in Sri Lanka before too long.

And so the crescendo mounts. The invisible lines of desire pull us toward predictable fates. In 2009 a pair of researchers announced that the consensus predictions for late-century temperatures were well outside the growing range

of our most important crops. Agricultural economist David Battisti, of the University of Washington, Seattle, told *Discovery News* that "the changes are so big and in the wrong direction in places where it really matters [that] if we don't adapt, we will have really serious problems." He concluded that the only options will be migration or death. Geographer Bill Easterling, of Pennsylvania State University in University Park, put it this way: "They've given us an early glimpse into territory beyond charted human experience."

And so traditional desire—if, that is, you count modernity as our tradition—has pushed us over the edge of a cliff. If we wish to cushion our fall, our desires will need to change.

Which is fine. Because the set of desires we have constructed have done more than damage the planet. They have also damaged us, turning us into new kinds of people who are radically disconnected from each other. Cheap fossil fuel has had three effects: it has made us rich, it has destroyed our climatic balance, and it has created the first race of people who have no practical need of their neighbors. Food—represented here as yellow crates of chicks and sinister rows of hanging tomatoes—comes from some distant and unknown place. Energy comes not from the muscles of your neighbors joining together to raise a barn or bring in a harvest but from some form of destruction. But forget for a moment the destruction of mountainsides or atmospheres and focus instead on the destruction of communities. In America, exemplar of this trend, 75 percent of people have no relationship with their next-door neighbor. The average American has half as many friends as the average American fifty years ago. We eat meals with friends and family half as often as we did fifty years ago. And not surprisingly, we are less satisfied: barely a quarter of Americans count themselves as "very happy" with their lives, despite the insane material surfeit with which we live.

We are social animals, though. How many generations ago were we all apes, sitting in a clearing and grooming each other all day long? The trip from that to our current disconnection has left us unhinged and dazed. The only solution—and this will be no surprise to anyone who has looked at any of our spiritual traditions or given the matter any sustained thought—is to replace one set of desires with another. Not stuff—relationships. Not money—neighbors. Not "recreation" or "leisure"—participation. Not ease—connection.

Consider the kinds of economy not shown in these photographs, the ebullient and human types increasingly buried under the weight of industrial consumerism. Most humans go shopping at small markets where farmers and artisans sell a variety of goods they have grown or made themselves. Does this seem old-fashioned? Perhaps, but it is, in fact, the fastest-growing part of the American food economy, with sales up 15 percent a year for the last decade. And the draw is not just the fresh food, not just environmental conscience. It's that the average shopper at a farmer's market has ten times as many conversations as the average shopper at a supermarket.

Those conversations are an invisible web, just as the waves of desire are invisible. But their consequences are very real: as communities are slowly rebuilt and reconnected, much becomes possible. The substitution of buses and trains for all those private cars. The substitution of apartments downtown for starter castles sown across the far reaches of the suburbs. The substitution of local joy for distant consumption.

Imagine what this book would look like in a world that works. No more of the monolithic, the central, the oversized, the out-of-scale, the inhuman. Instead, the myriad, the diverse, the scattered, the low-powered, the well-attended. People in most of the pictures, and wild animals in the others.

It is sentimental to imagine such a world without working to build it. And many movements have arisen in recent years, movements that at their spear point form a movement to take the world seriously. For instance, the mission of 350.org, which I have helped birth, is to spread that threshold number, that shard of reality, as far and wide as humanly possible, to say that this number marks the edge of the cliff that we have gone over and if we work with great devotion, it is just possible we might get back up to the edge someday. Not before grievous damage has been done—the possibilities for a happy ending have long since passed—but while some possibility for human flourishing remains.

Such a movement needs to seek reality-based change, altering the economic signal that for two hundred years has driven our destruction, the one that says fossil energy is cheap and, hence, the first resort. We need to make fossil fuel expensive, so that we will use far less, so that we will turn to other sources, and so that we will live smaller, not traveling the globe, not building giant casino cities in the desert, not driving by ourselves in vehicles designed for forest rangers. The political challenge should not be underestimated. We are talking about transforming the desire system of the Western world, a desire system now spreading fast to every other corner of the earth. (Where it is still innocent: no one can stand along the Bund in Shanghai and condemn the Chinese who flock there to admire the skyline. There it still seems like hope .)

It is quite likely that we cannot pull it off, not in the short time science has given us. But how useful to be reminded, always, of the stakes. To see not lines on a graph but the calamitous reality of the world brought about by our desires. We need this not to be our legacy.

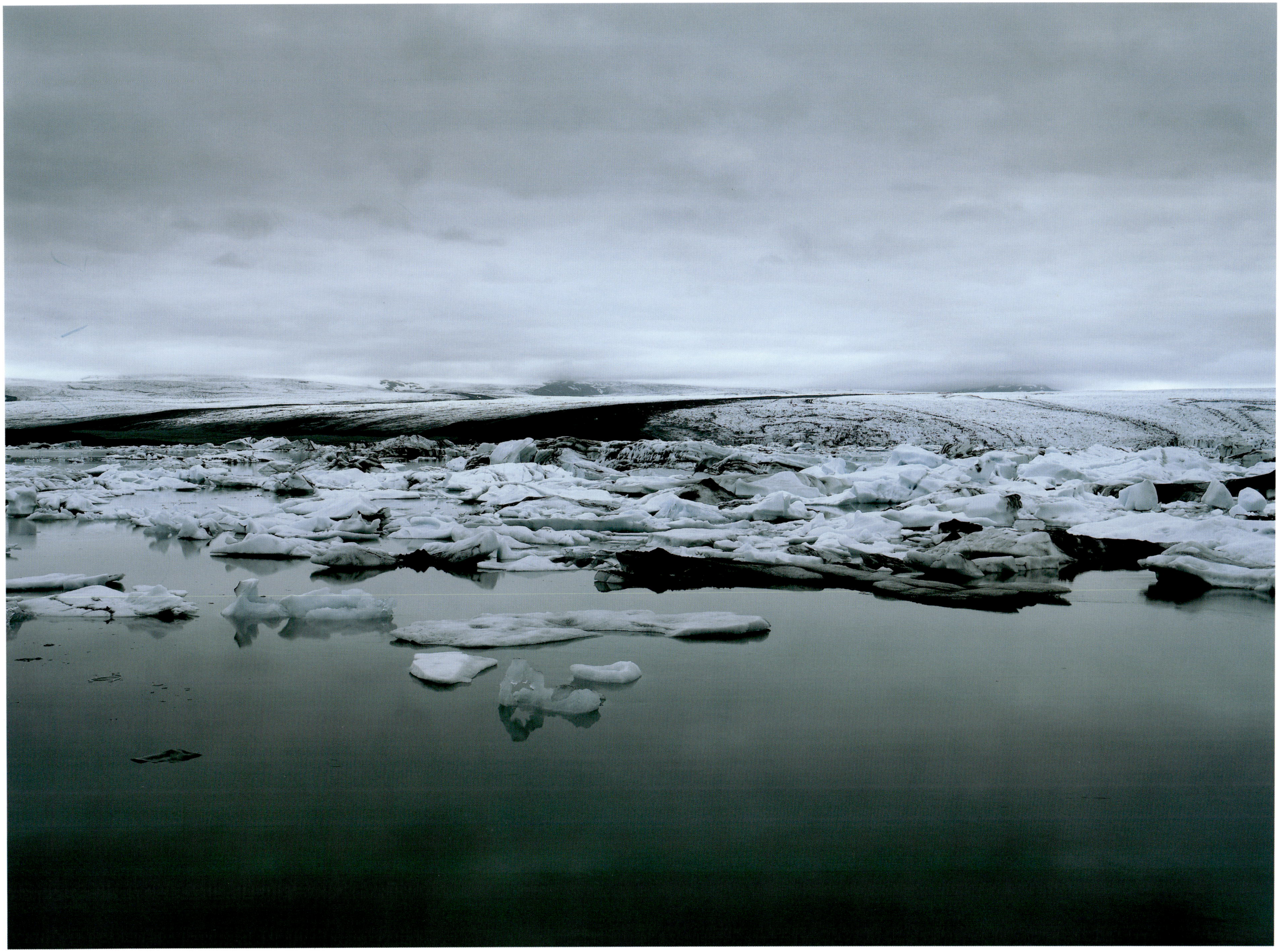

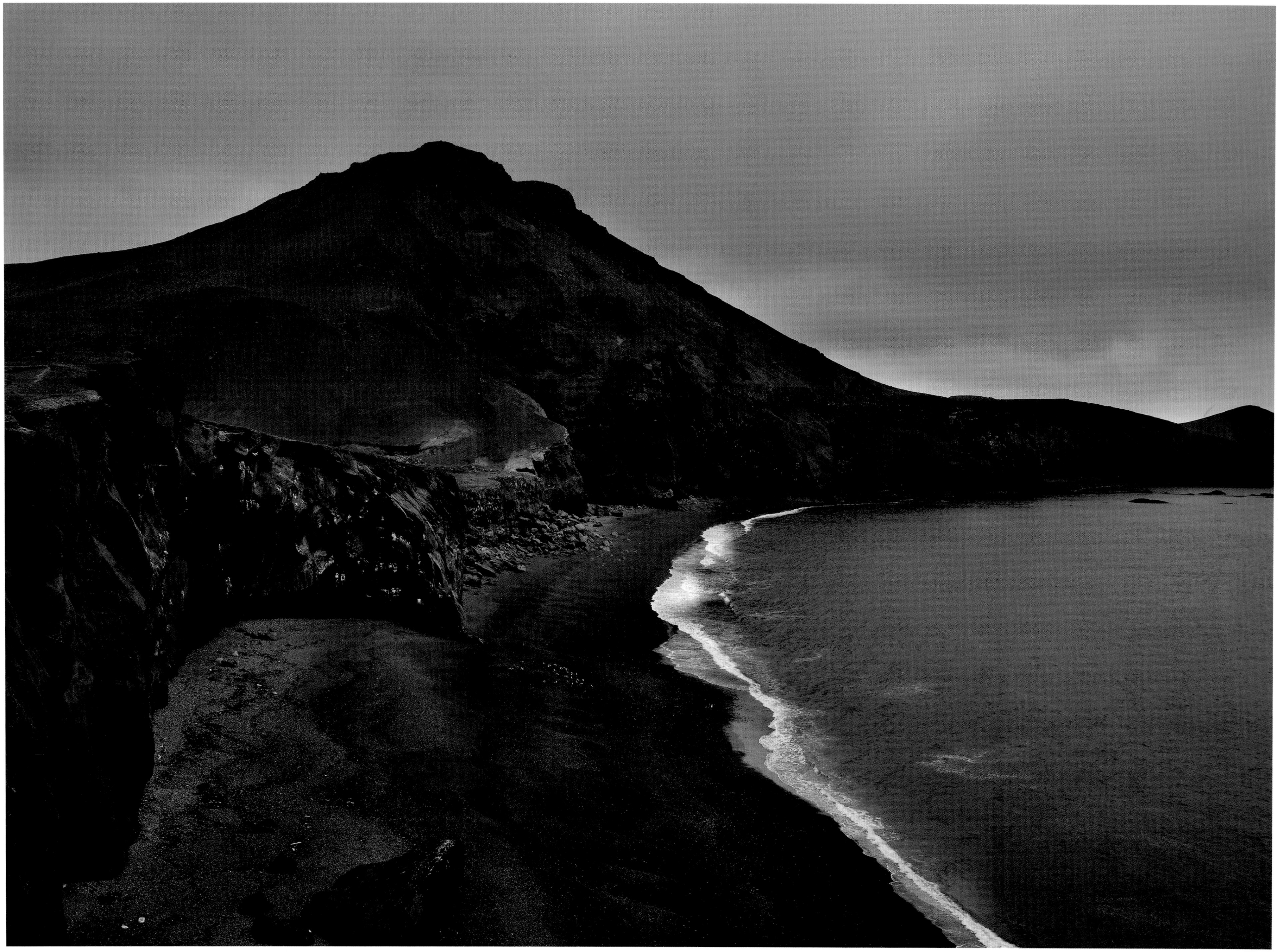

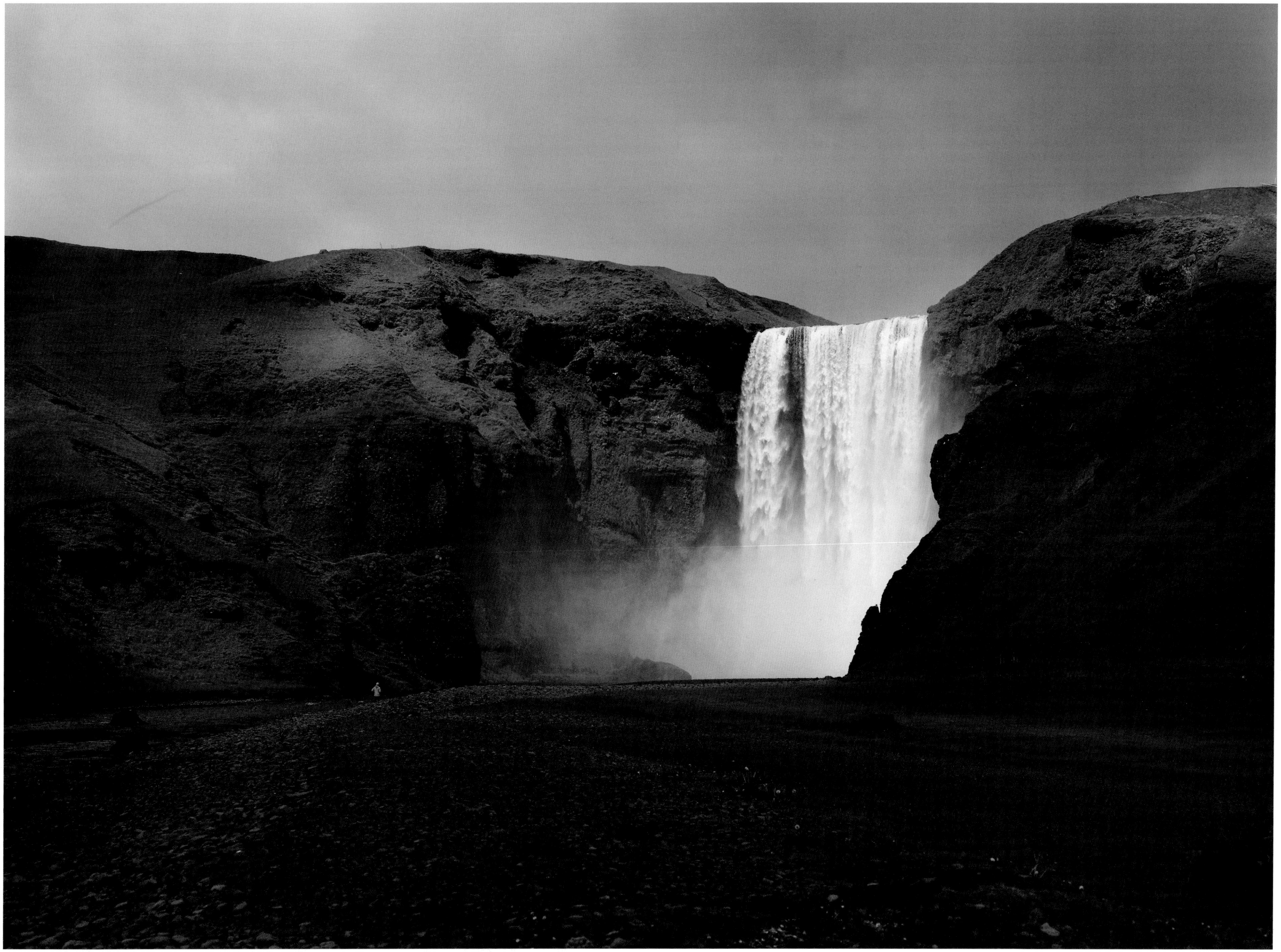

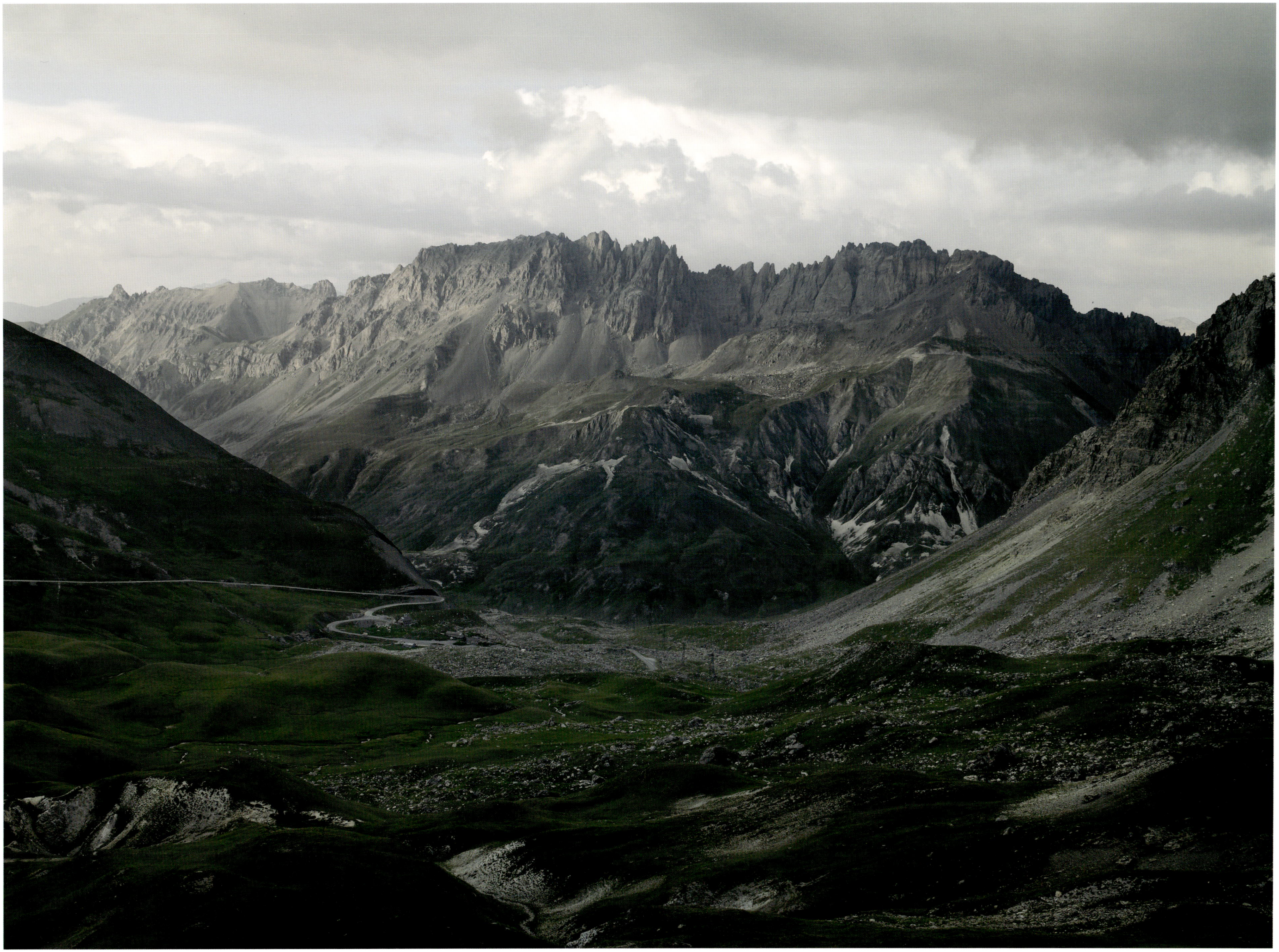

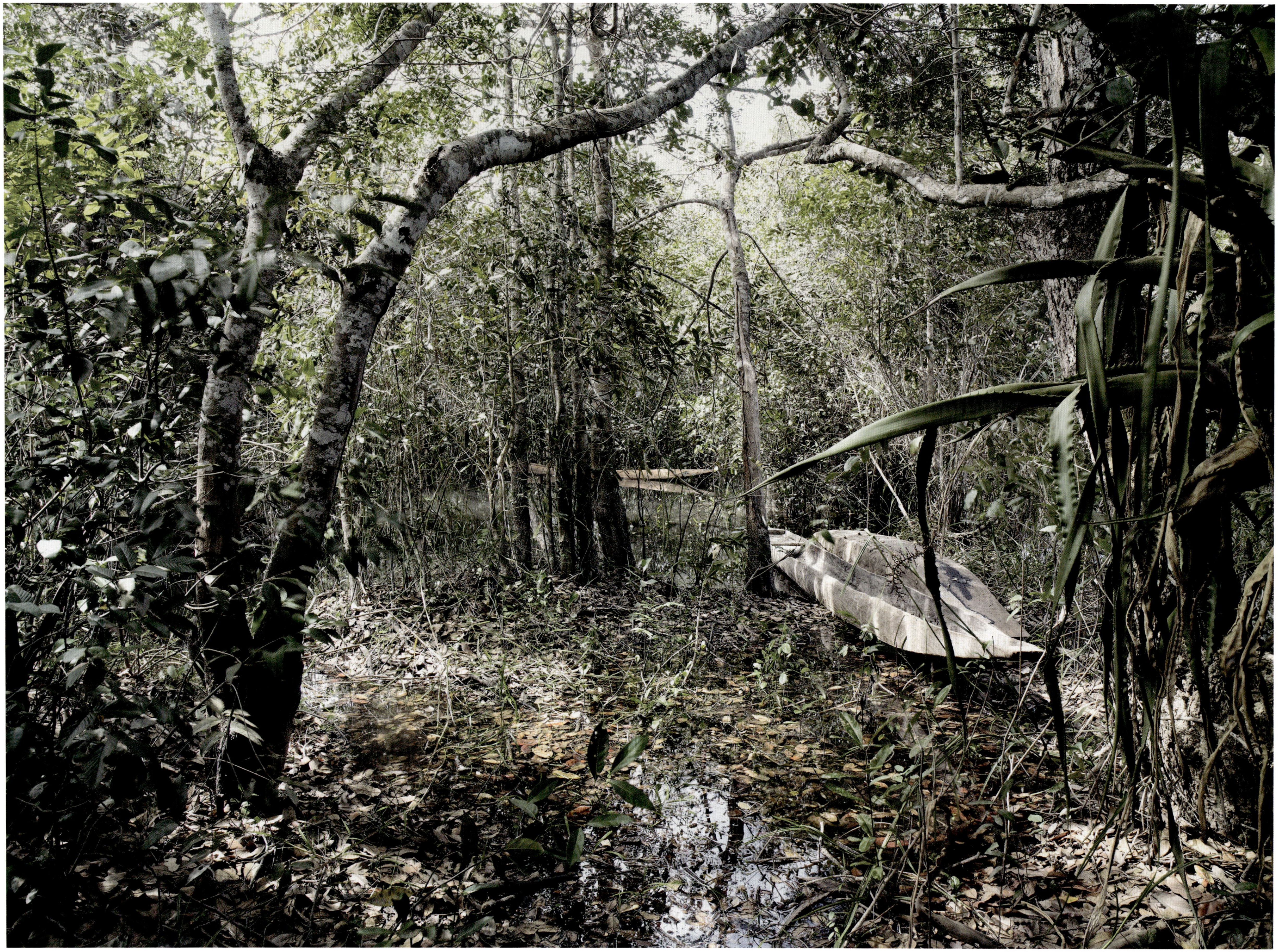

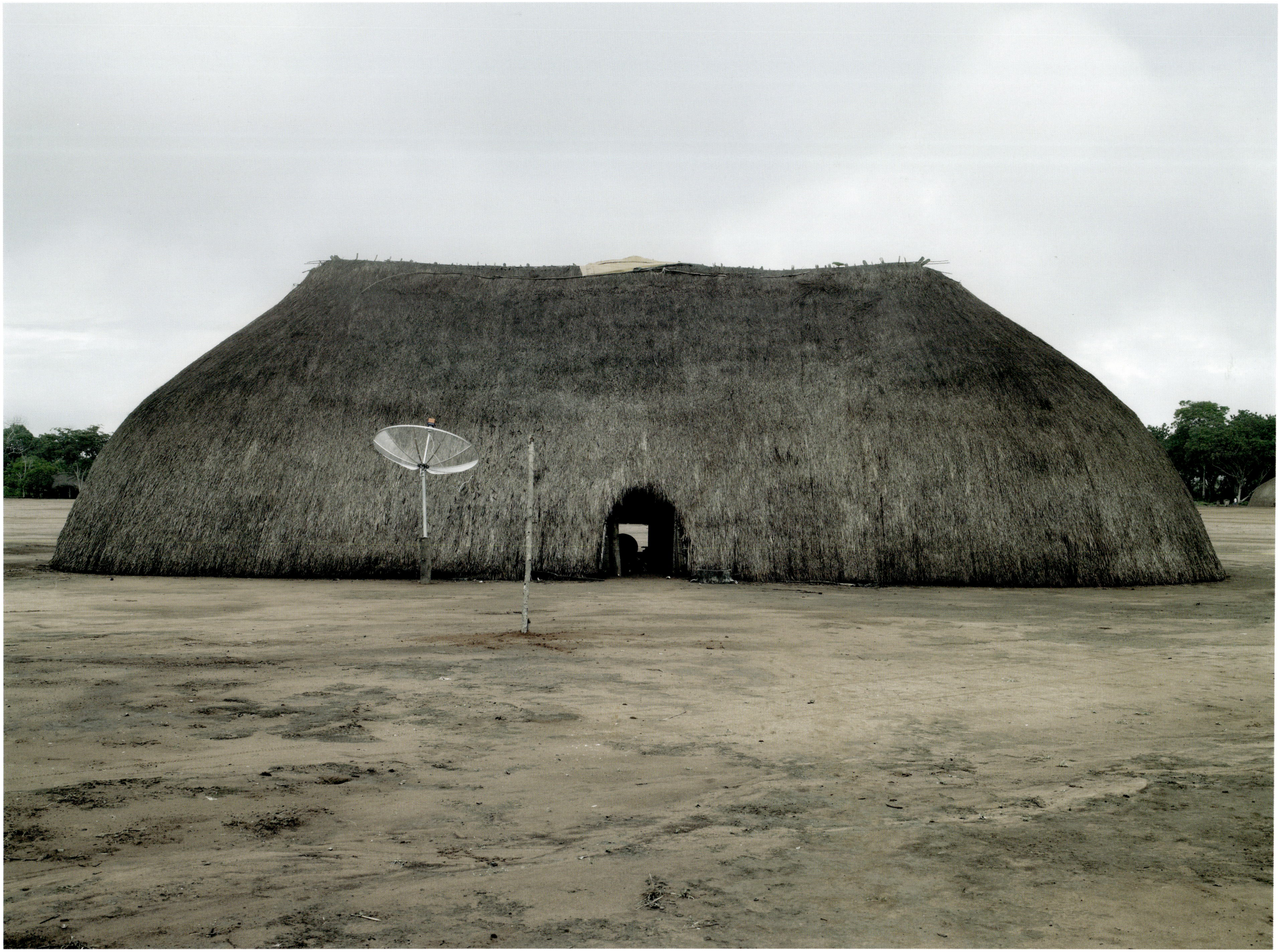

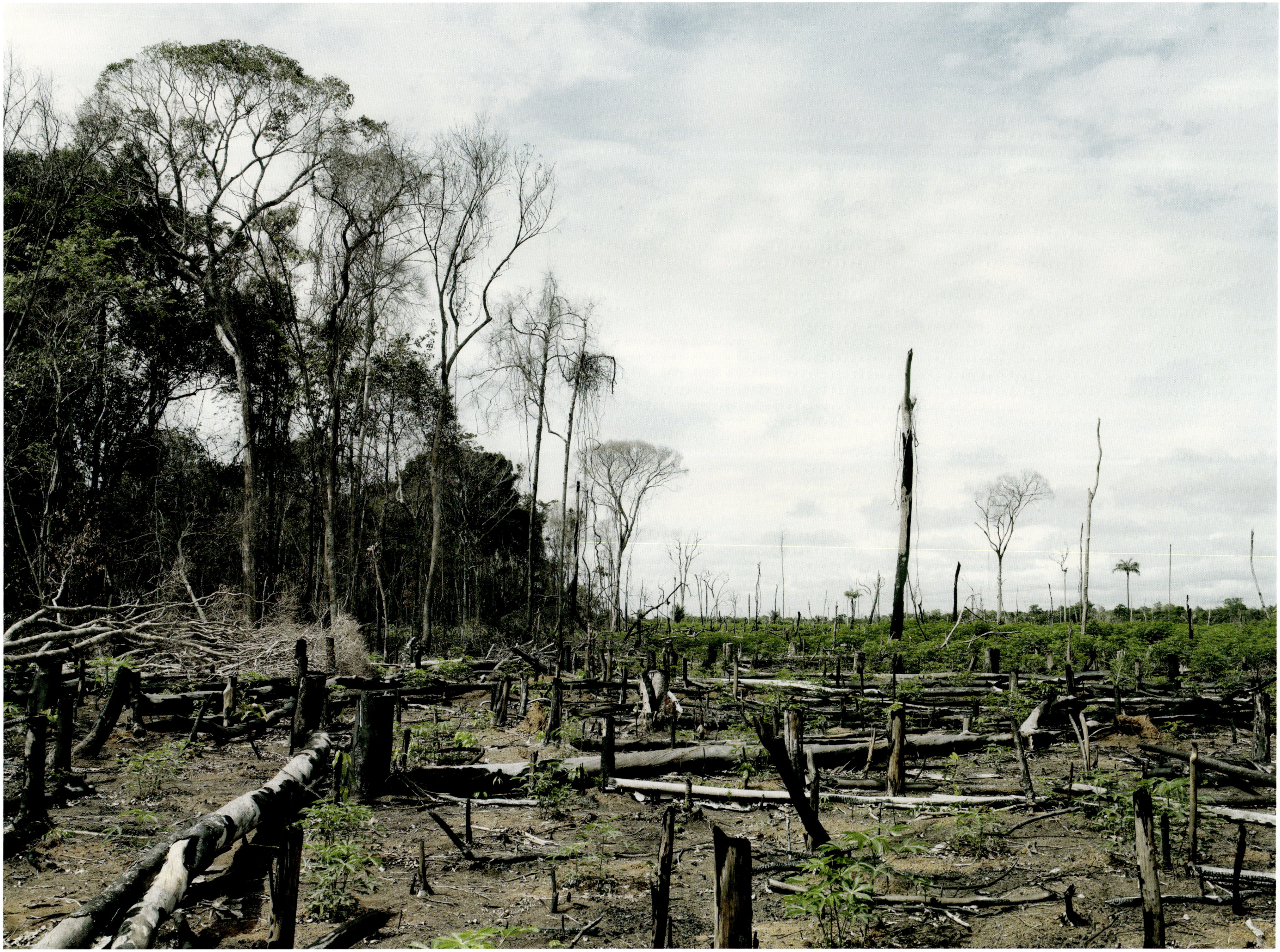

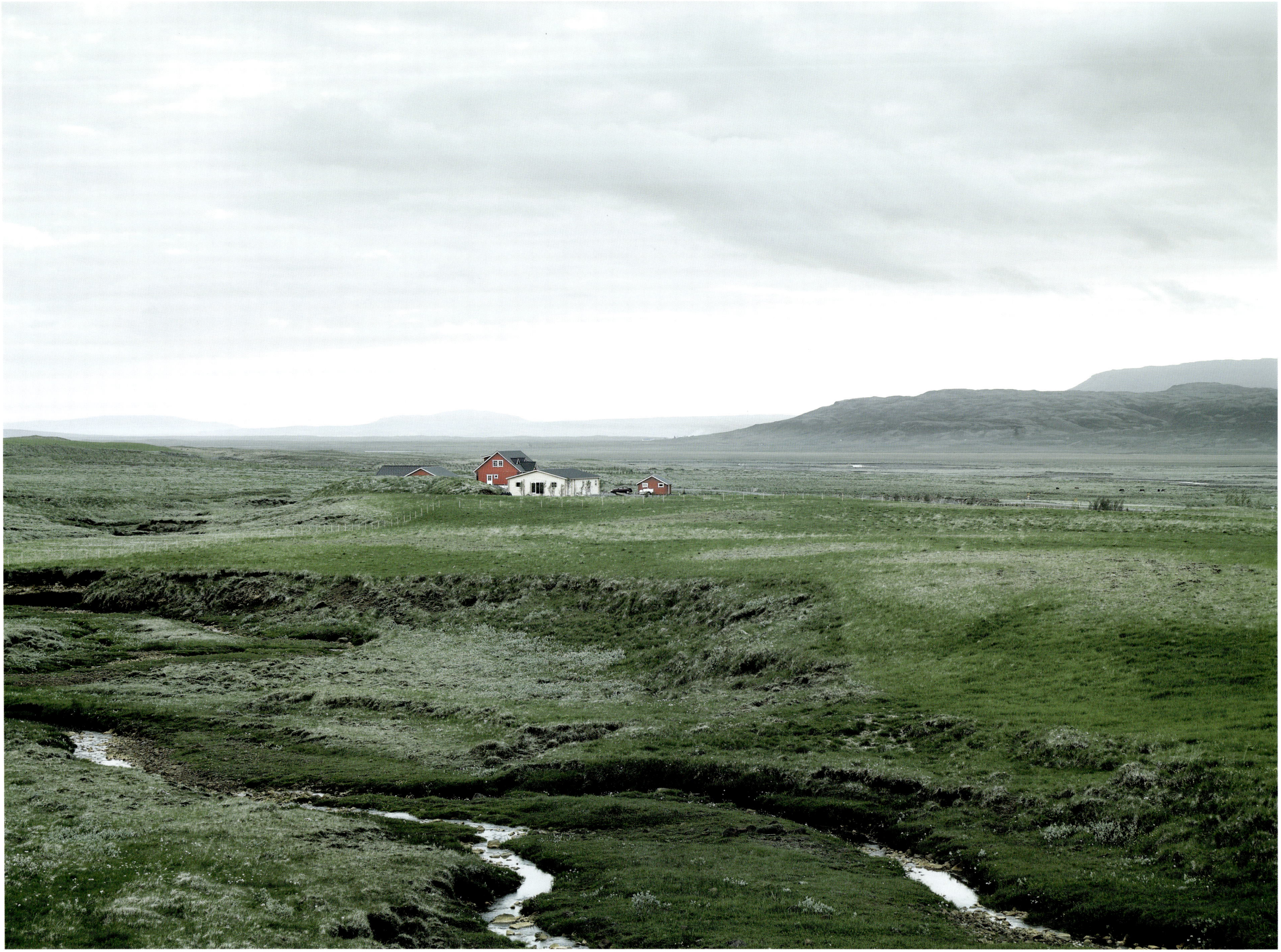

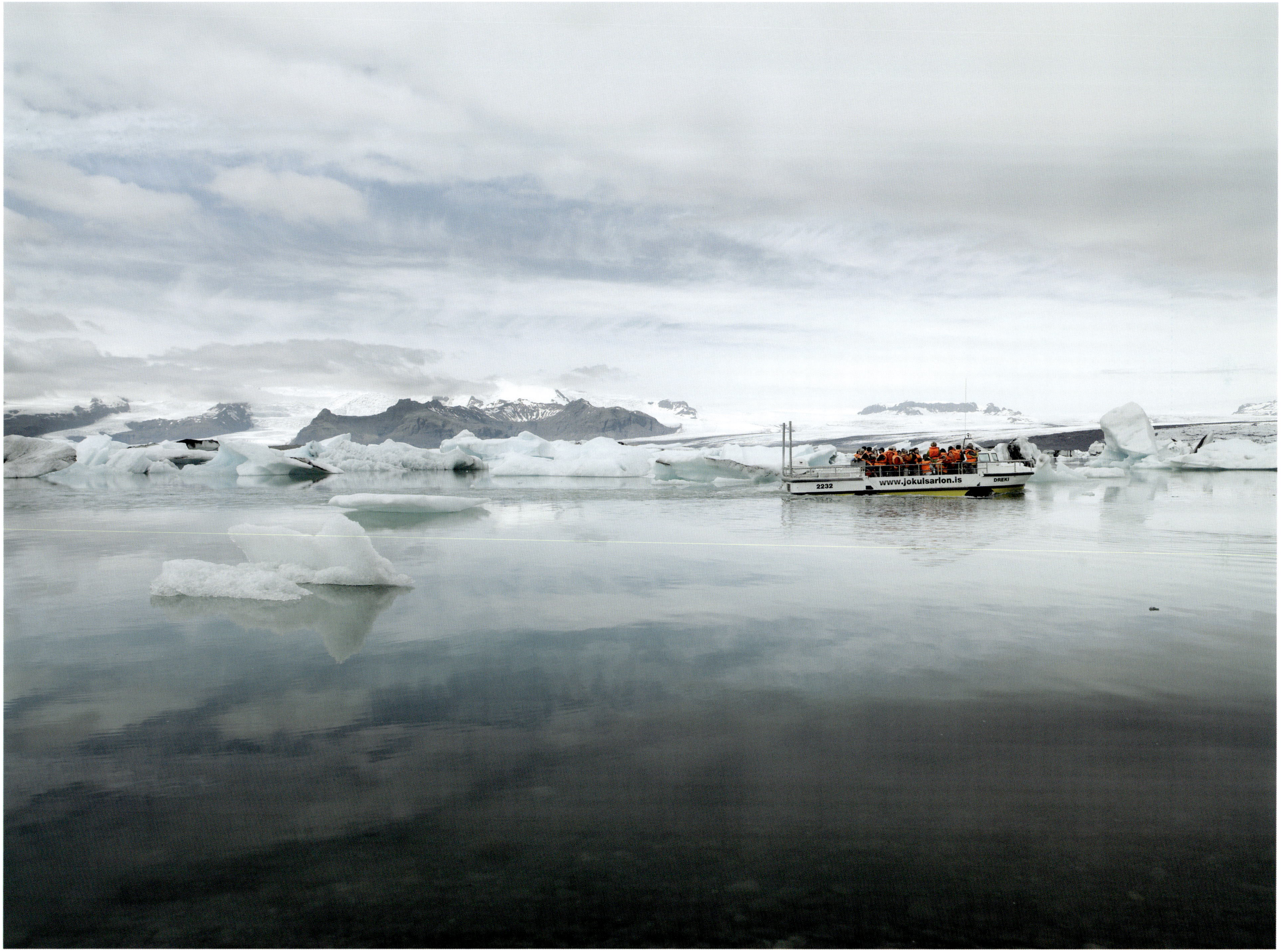
2232
www.jokulsarlon.is
DREKI

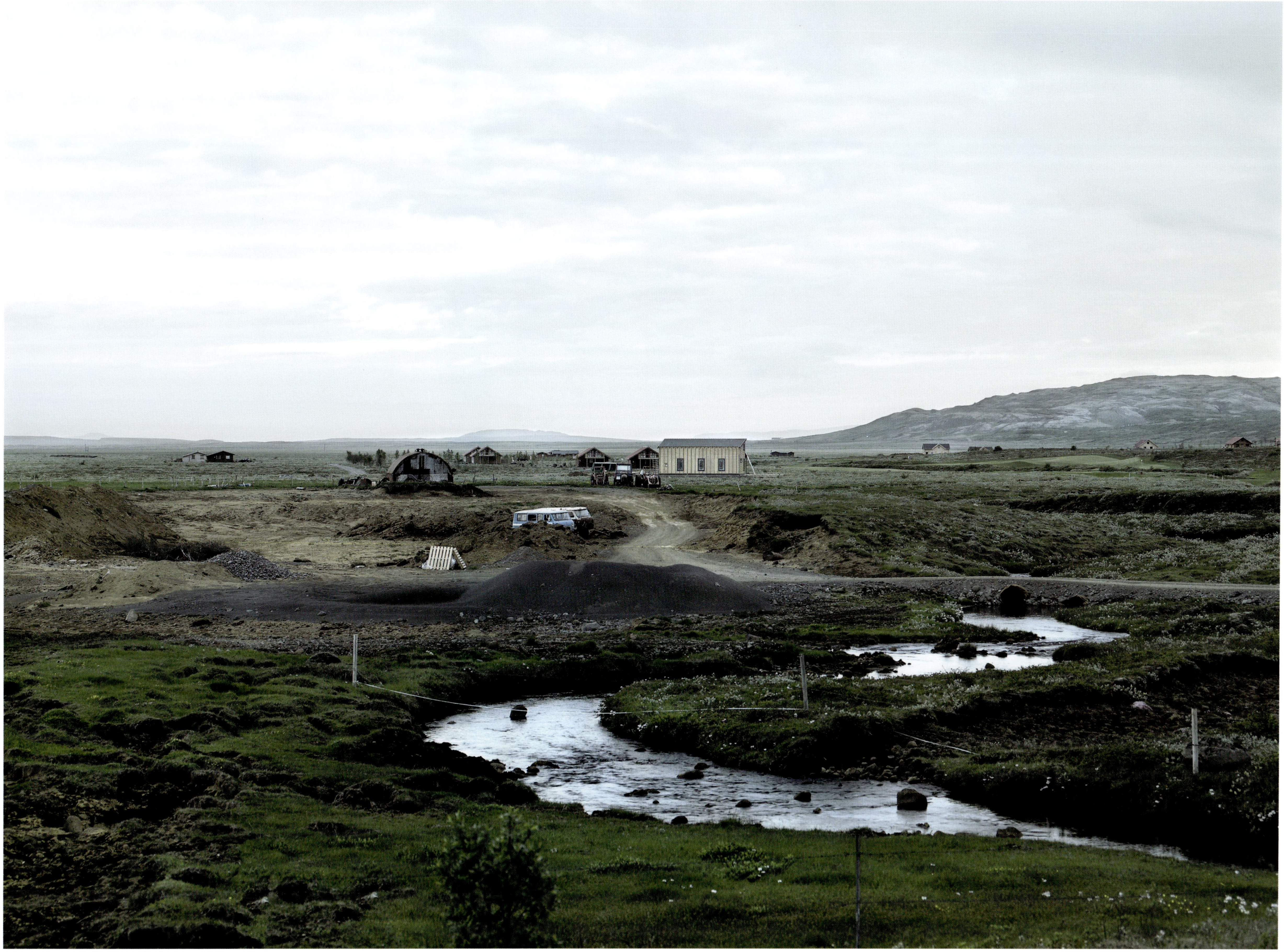

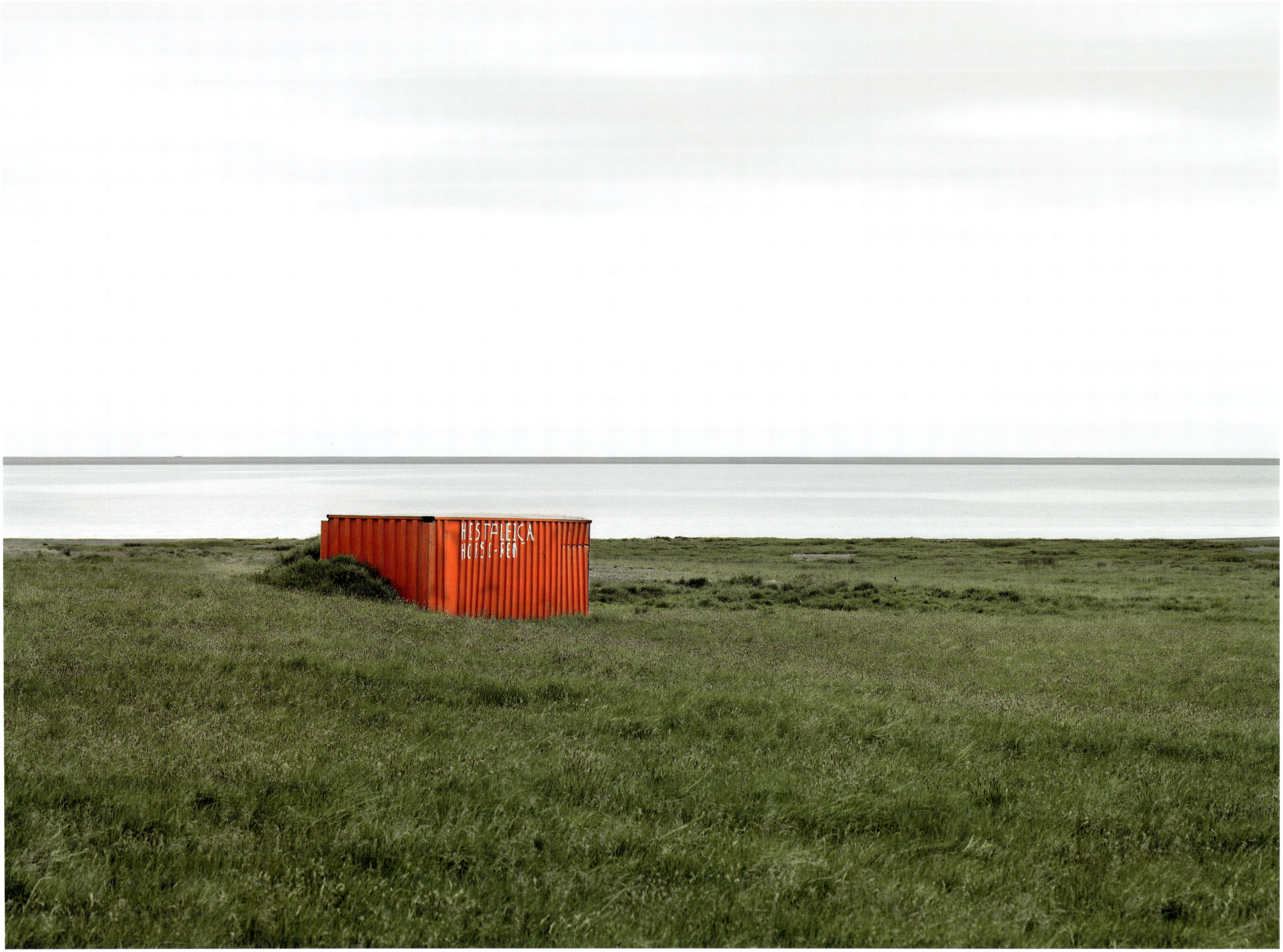
HESTALEIGA
HORSE-REN

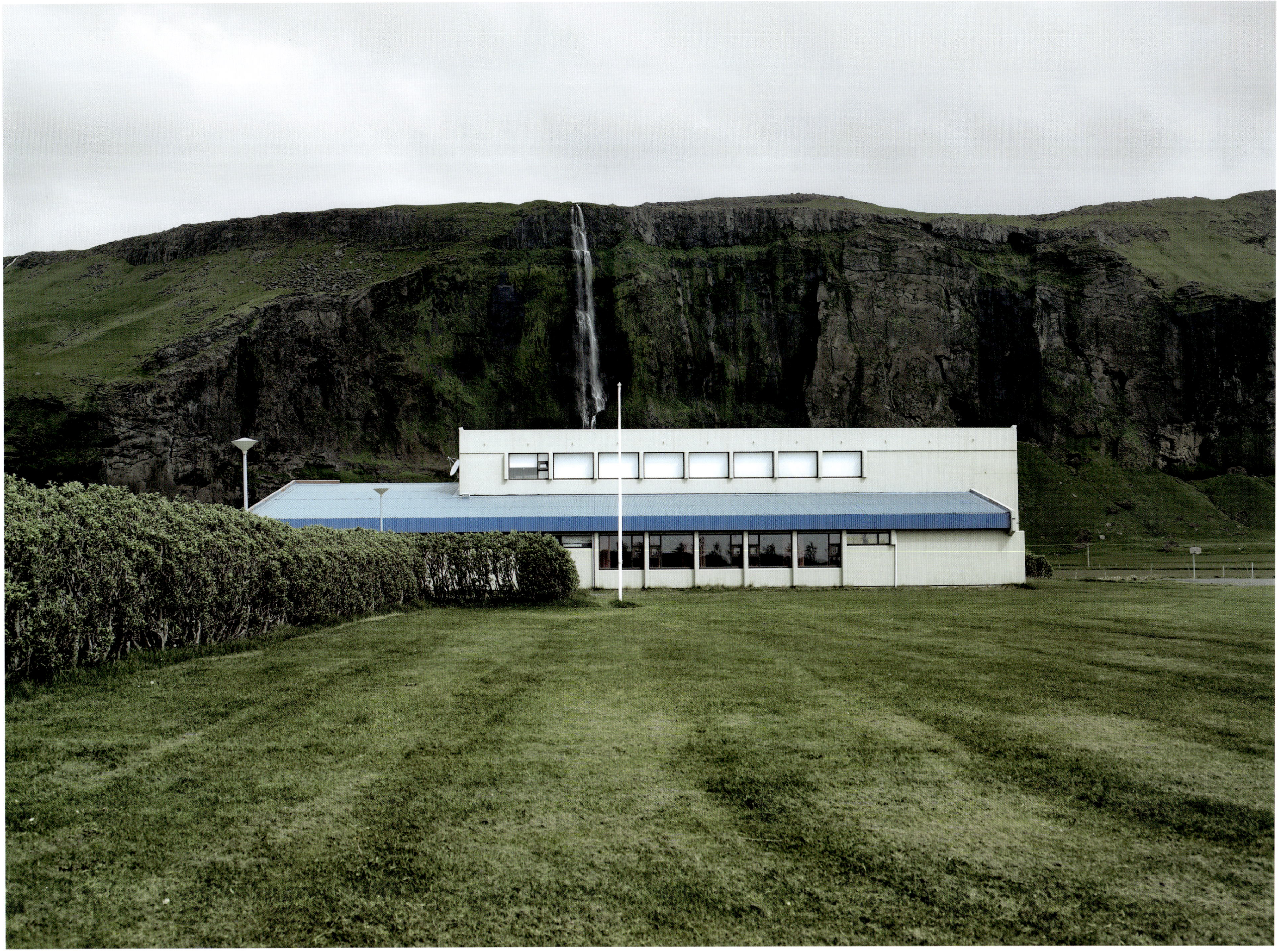

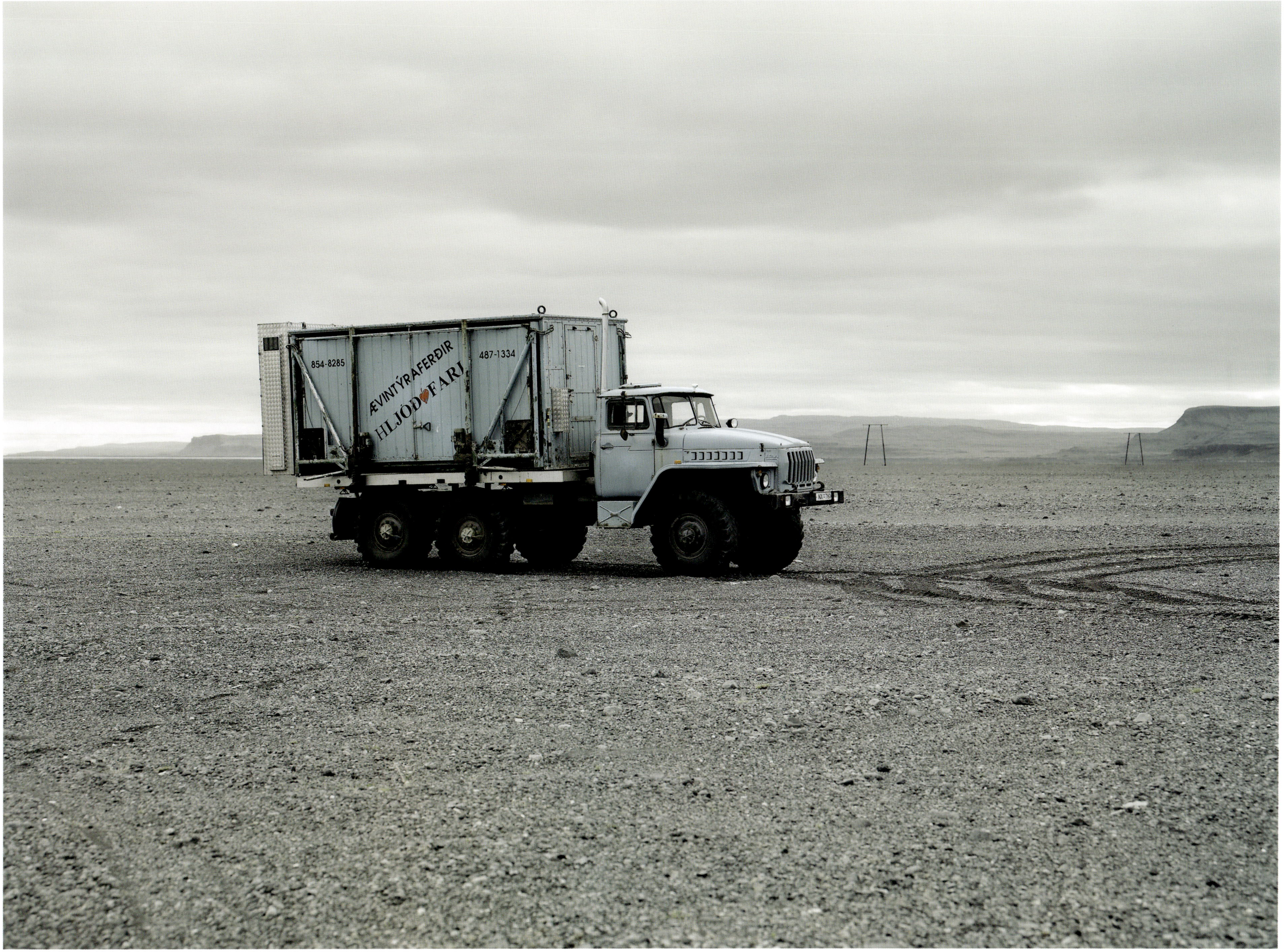
854-8285
ÆVINTÝRAFERÐIR
487-1334

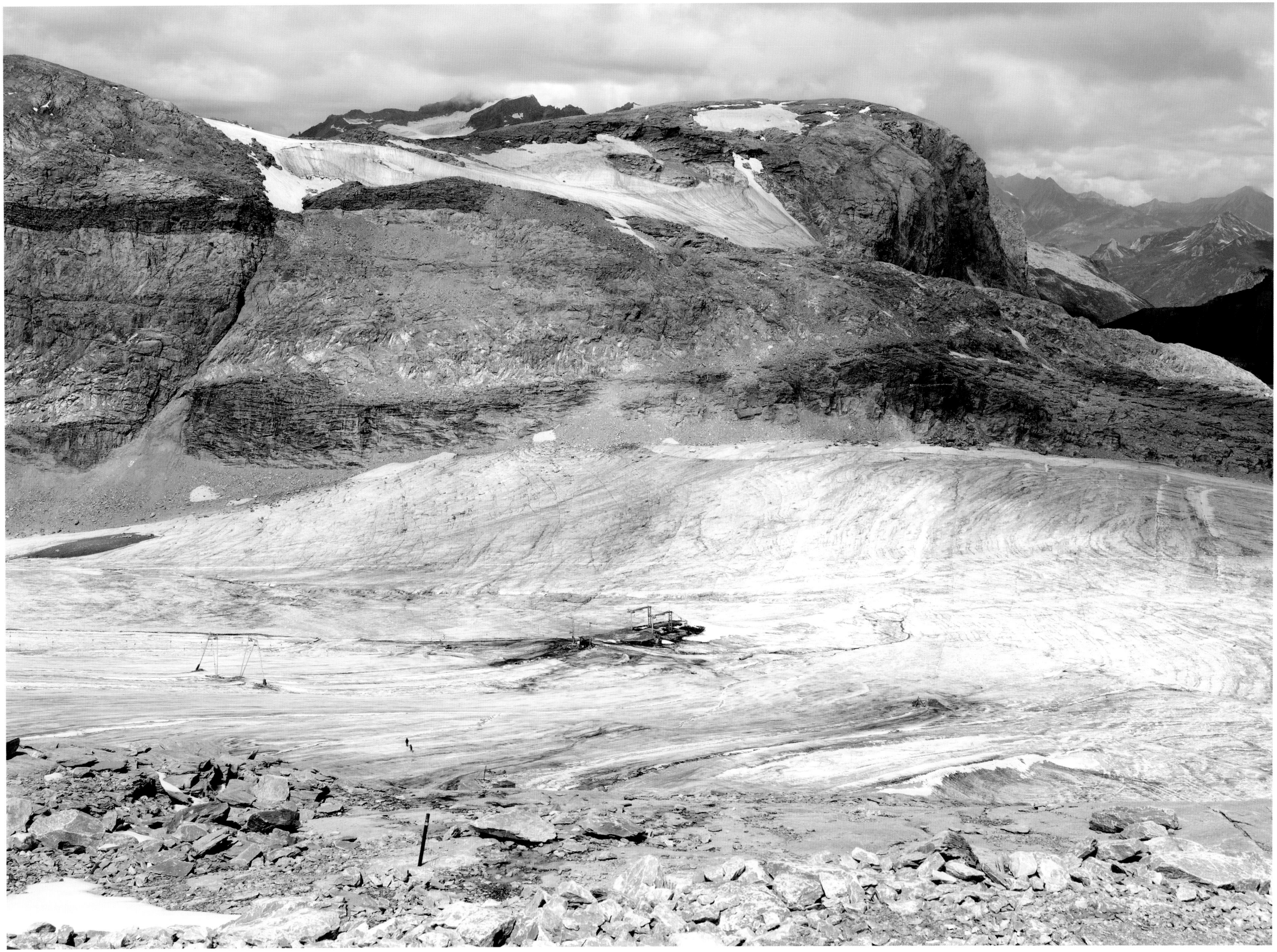

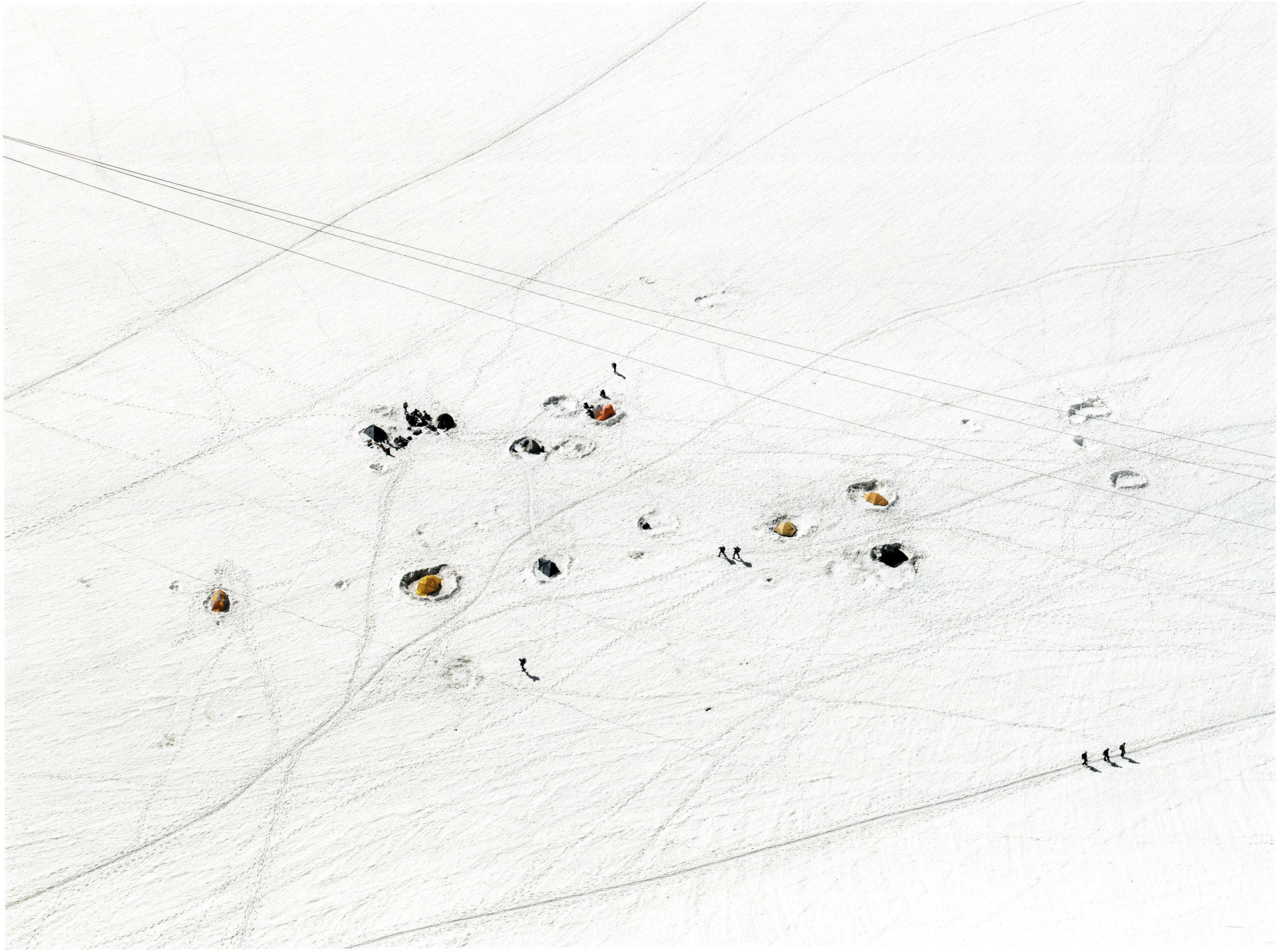

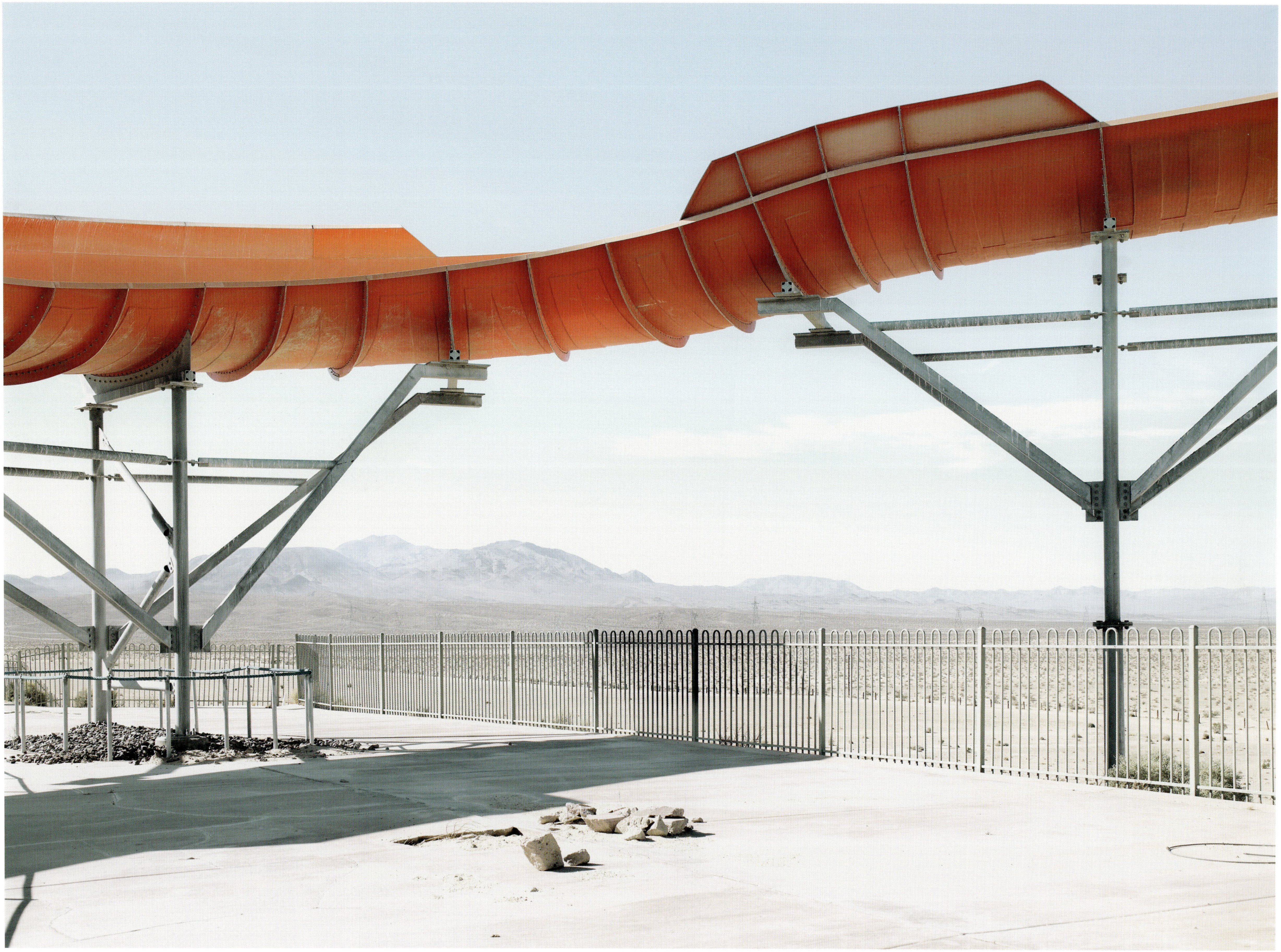

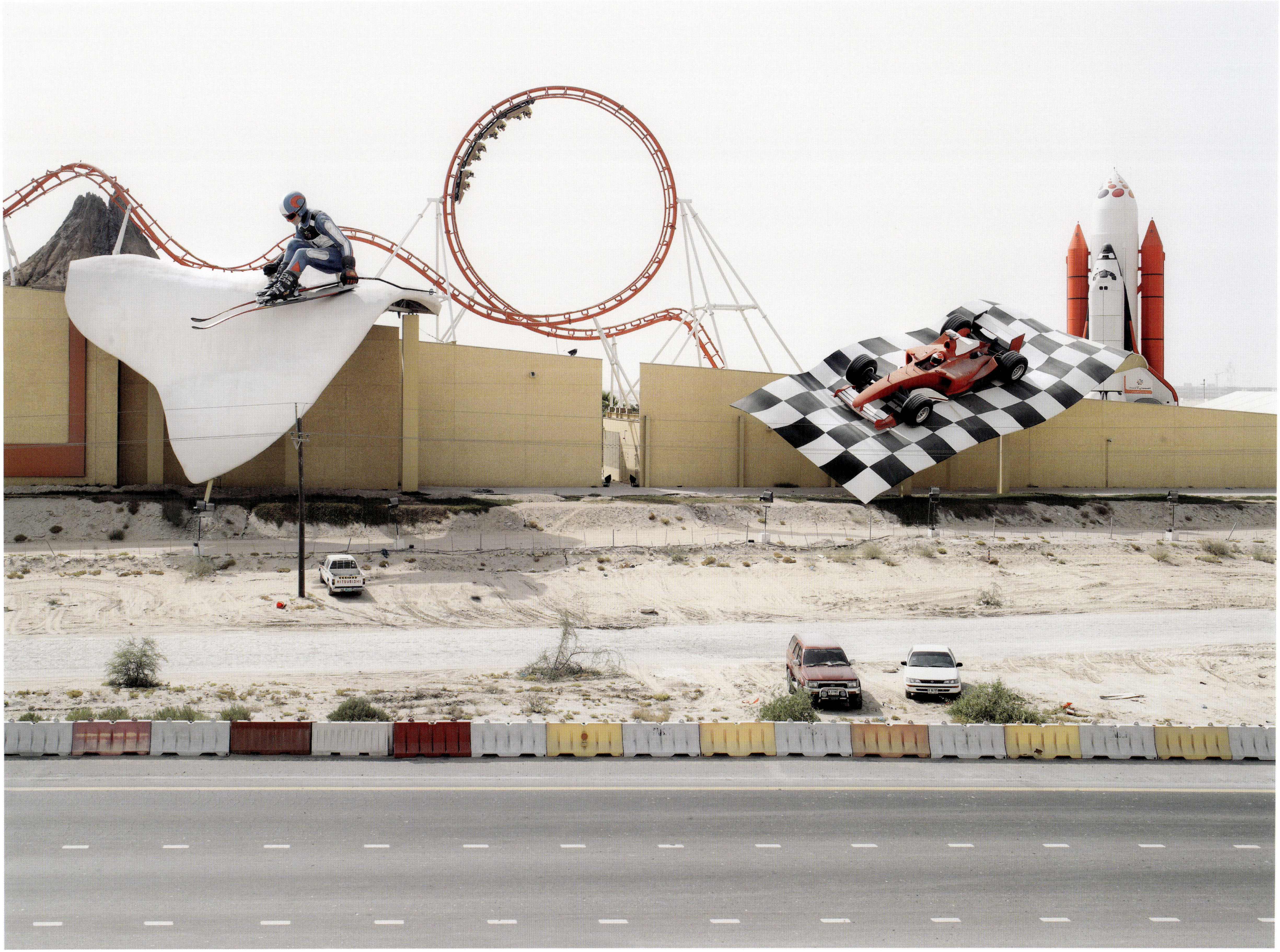

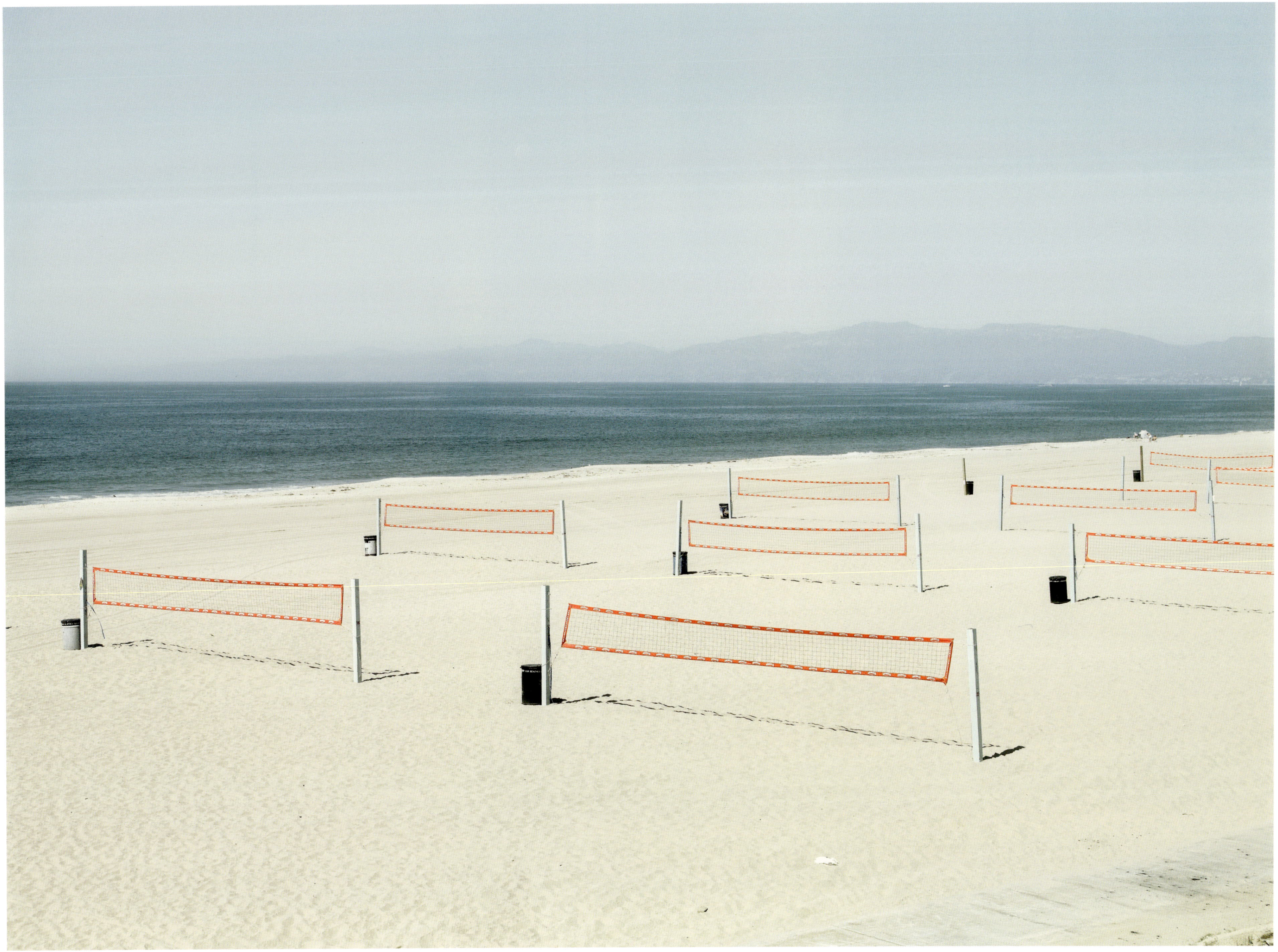

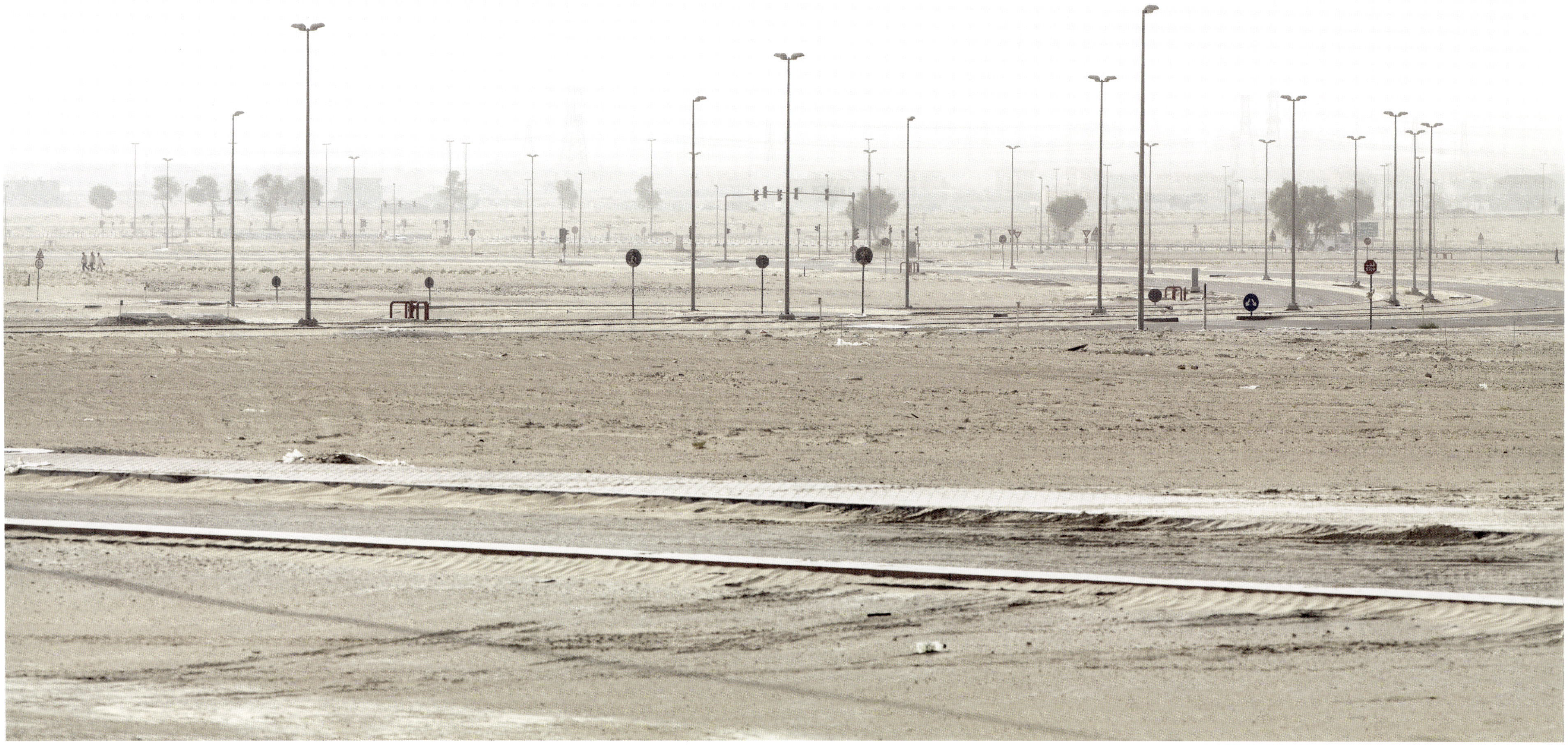

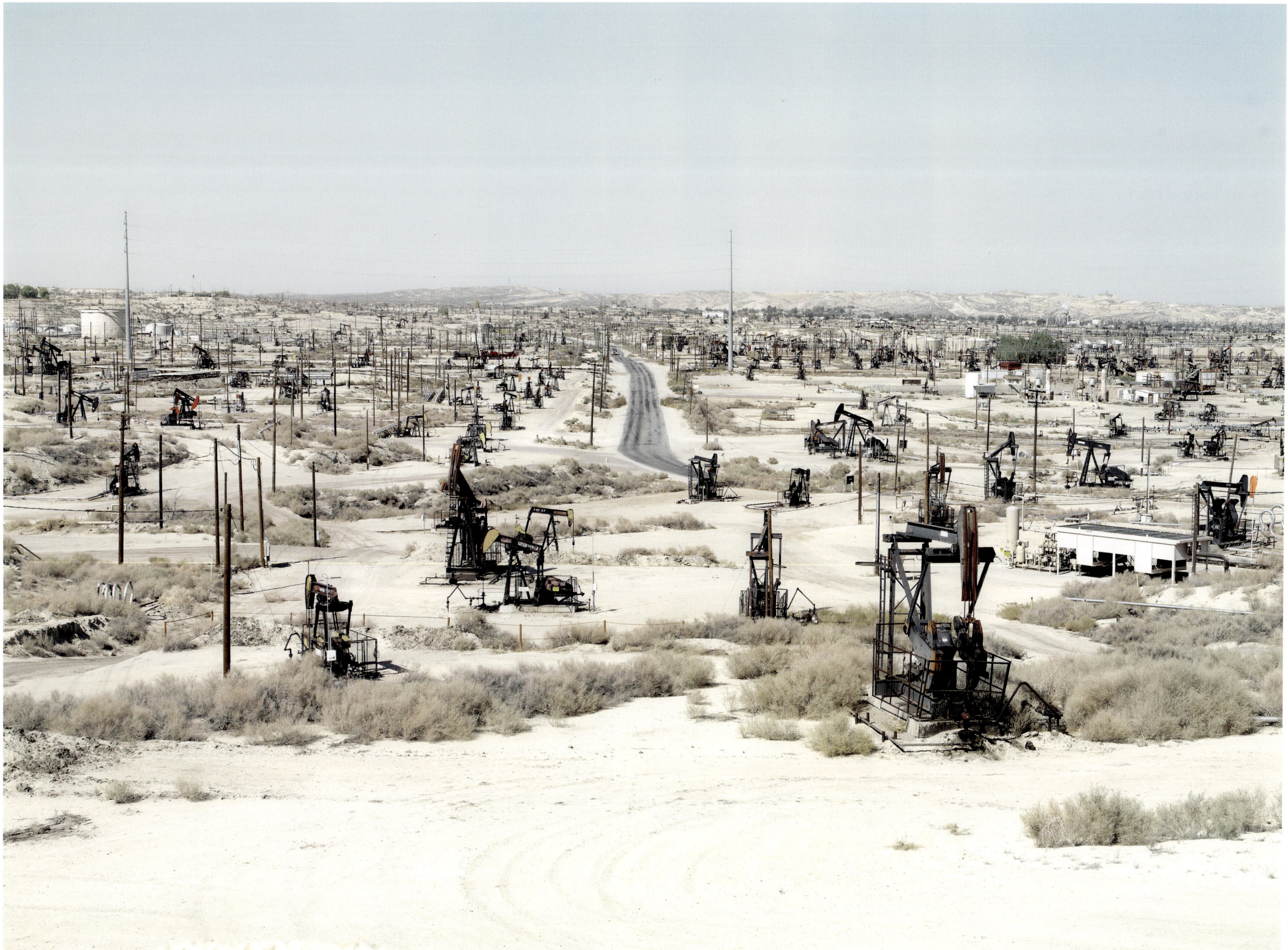

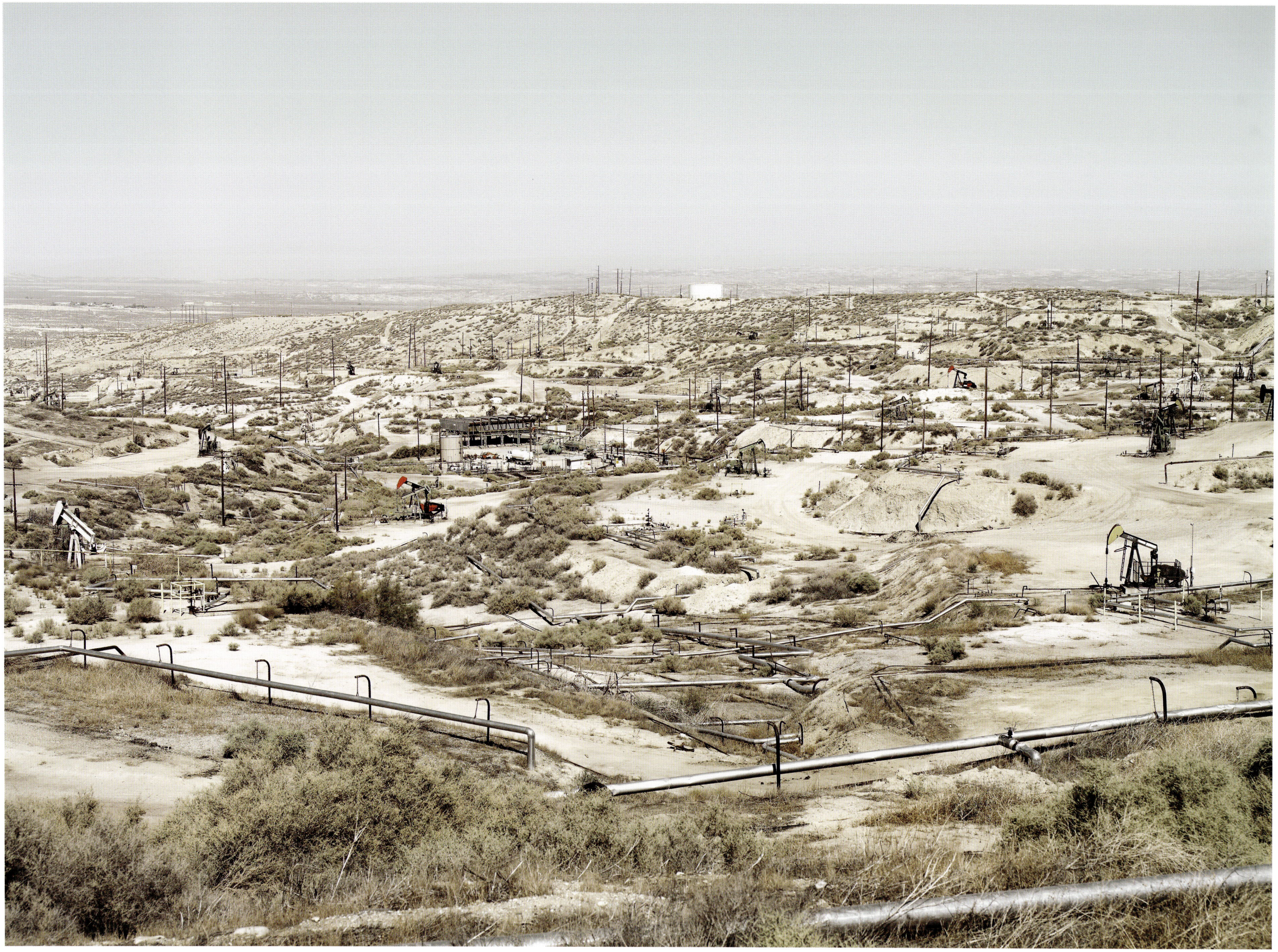

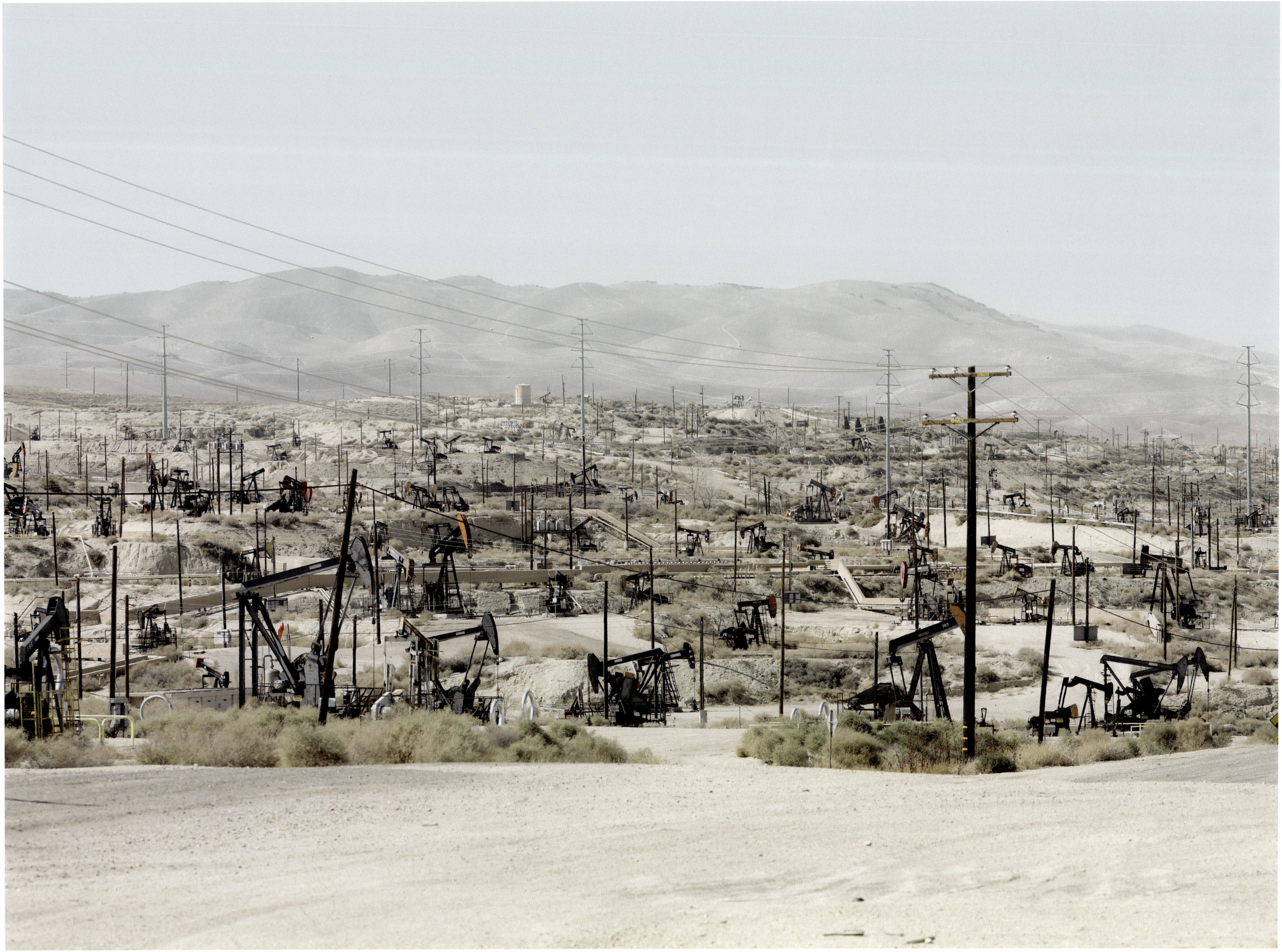

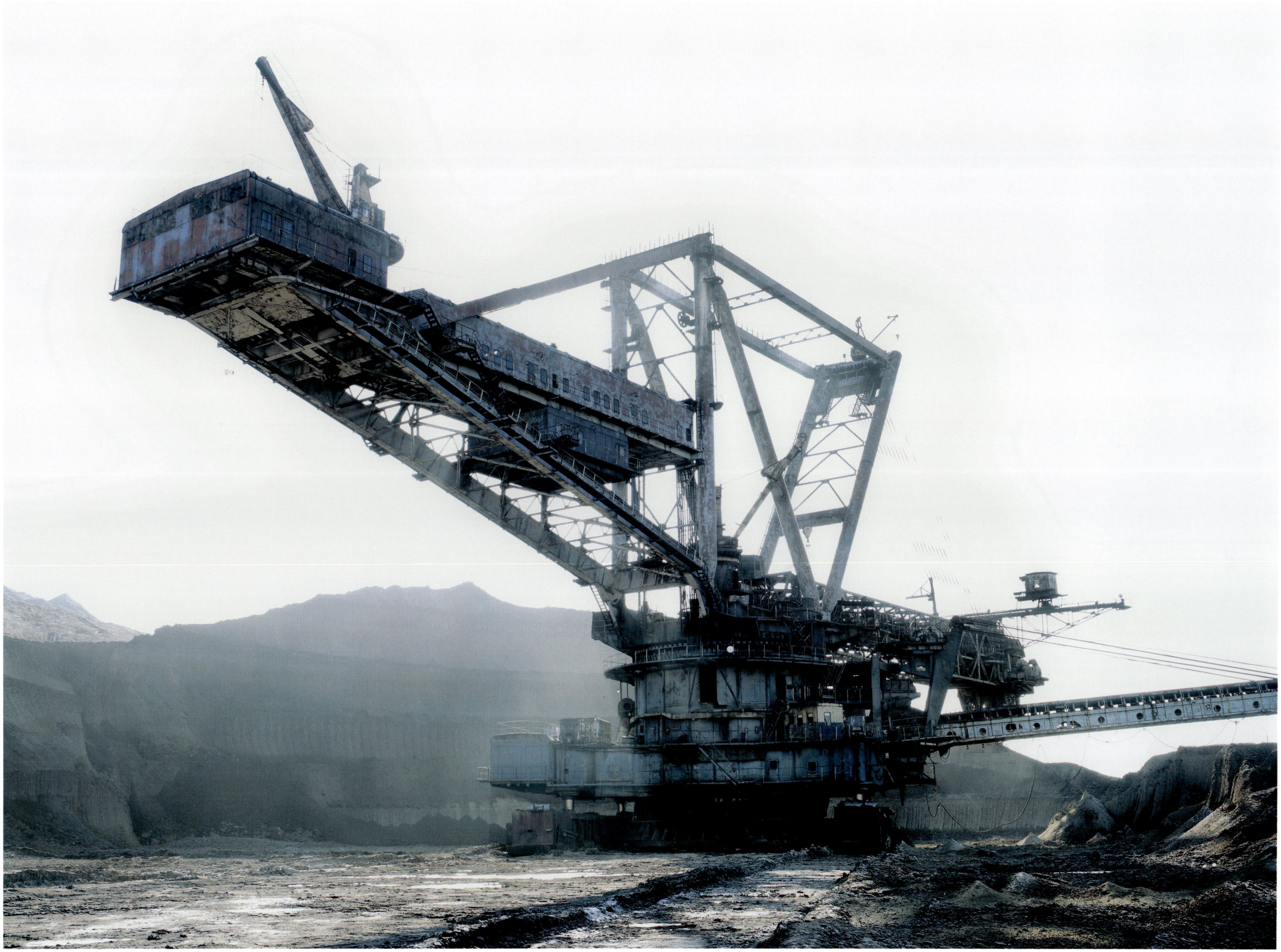

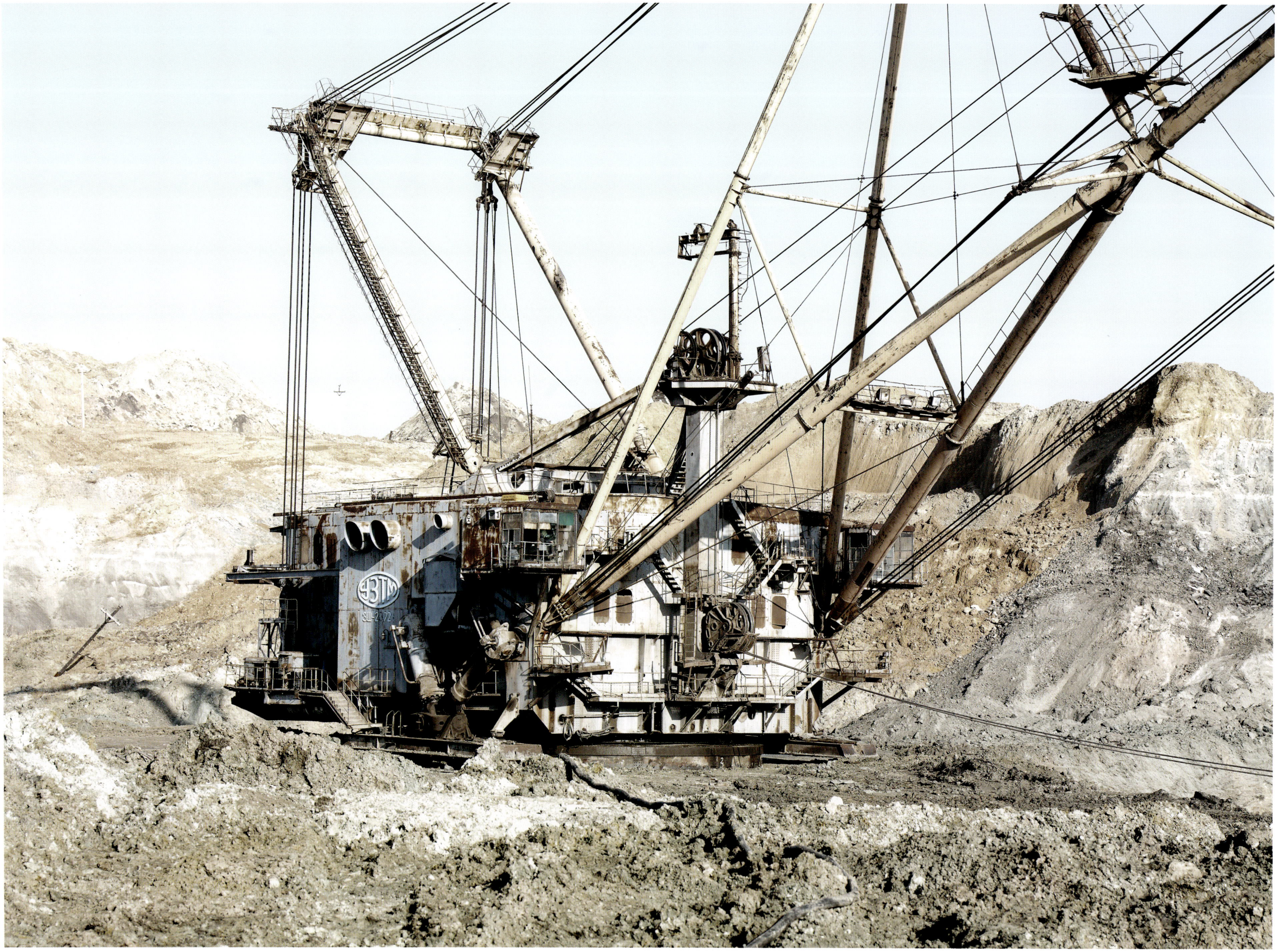
УЗТМ

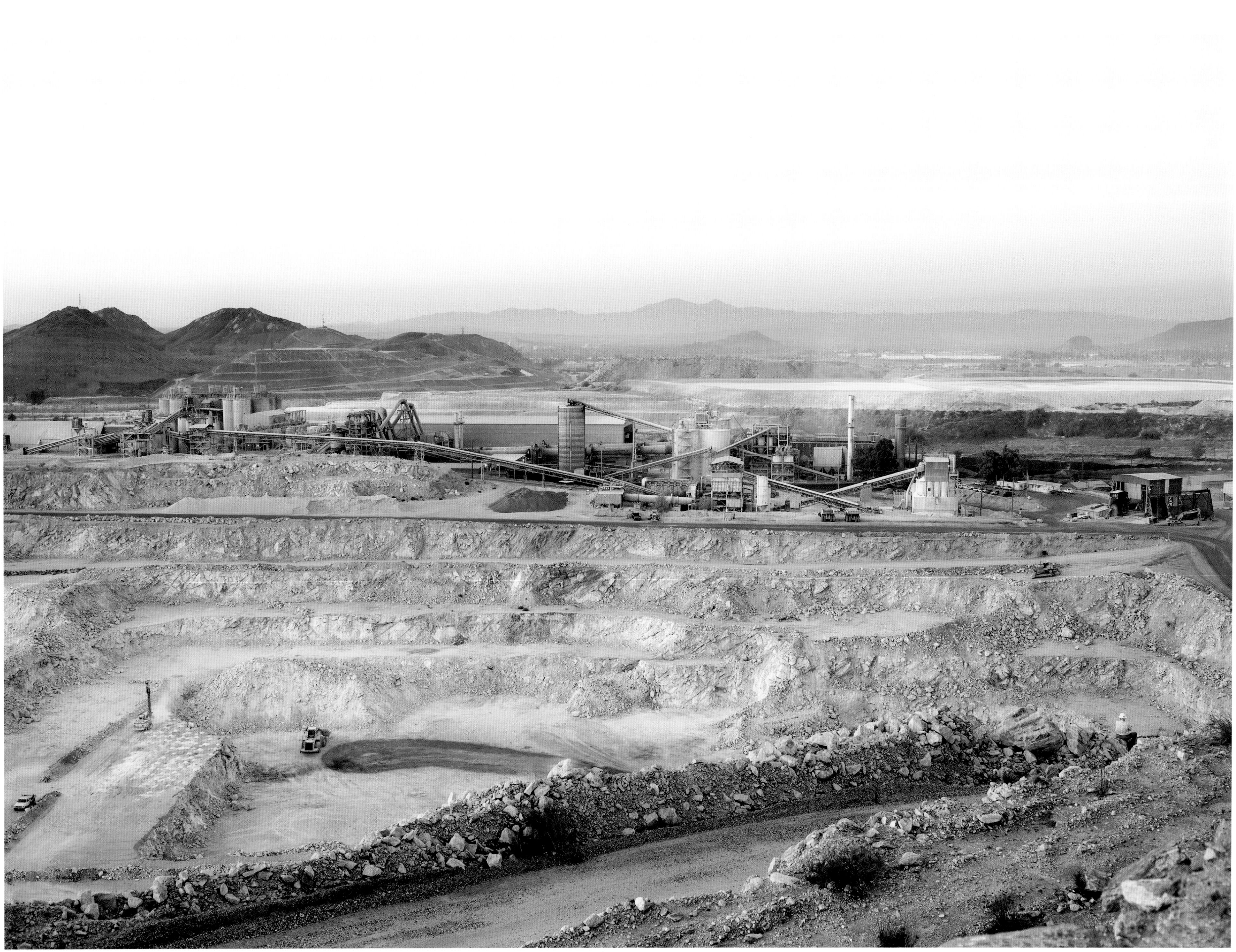

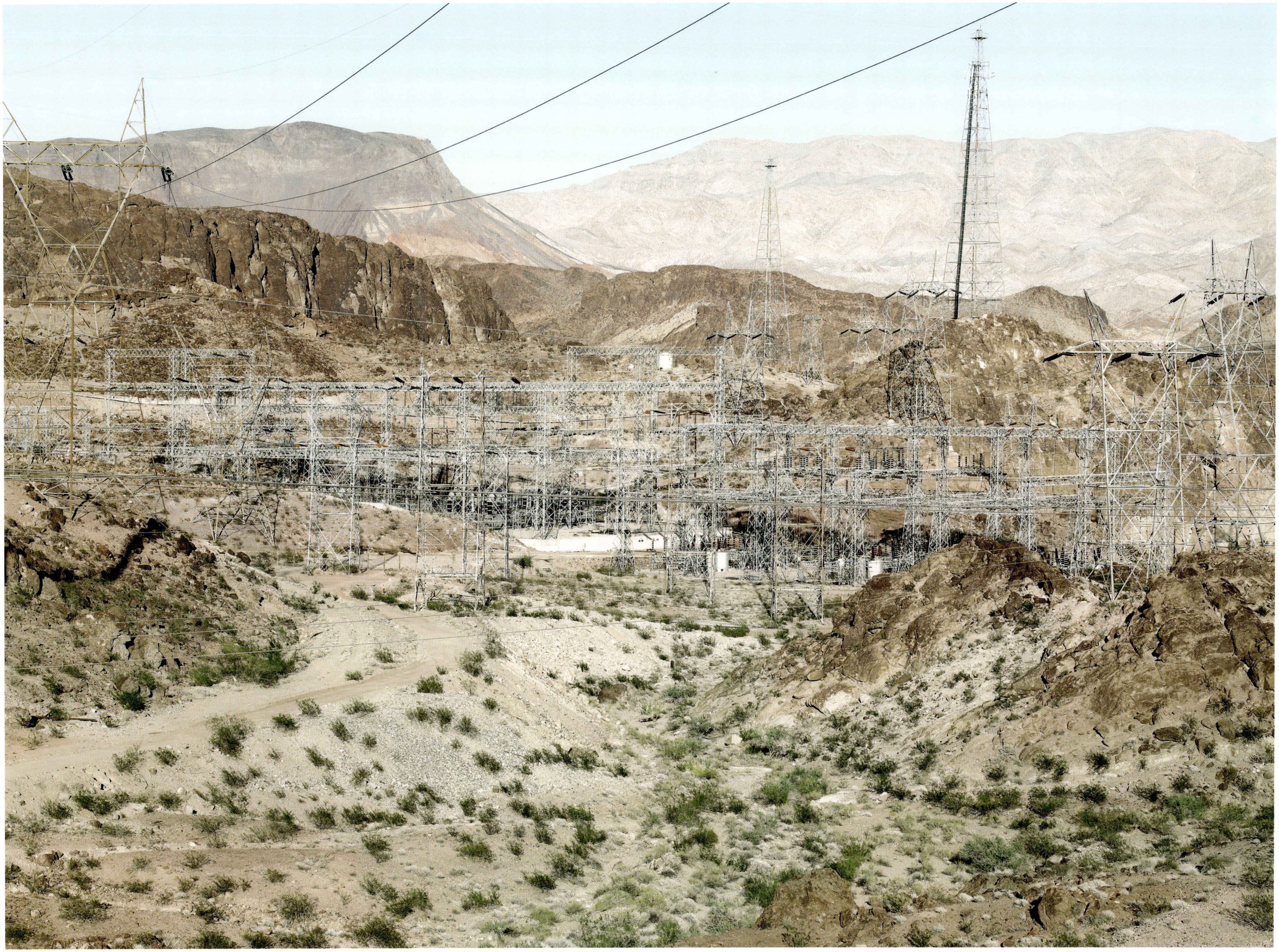

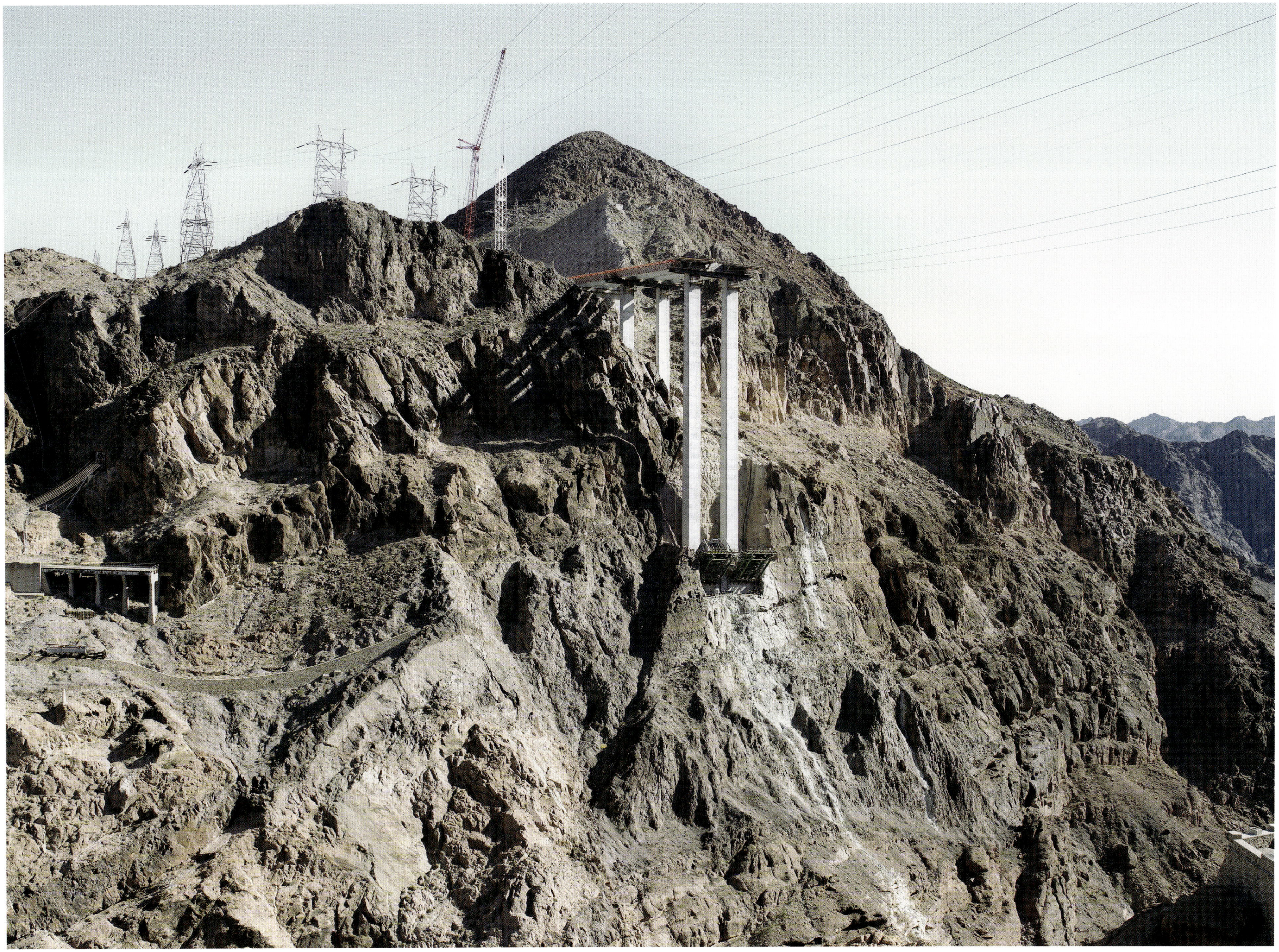

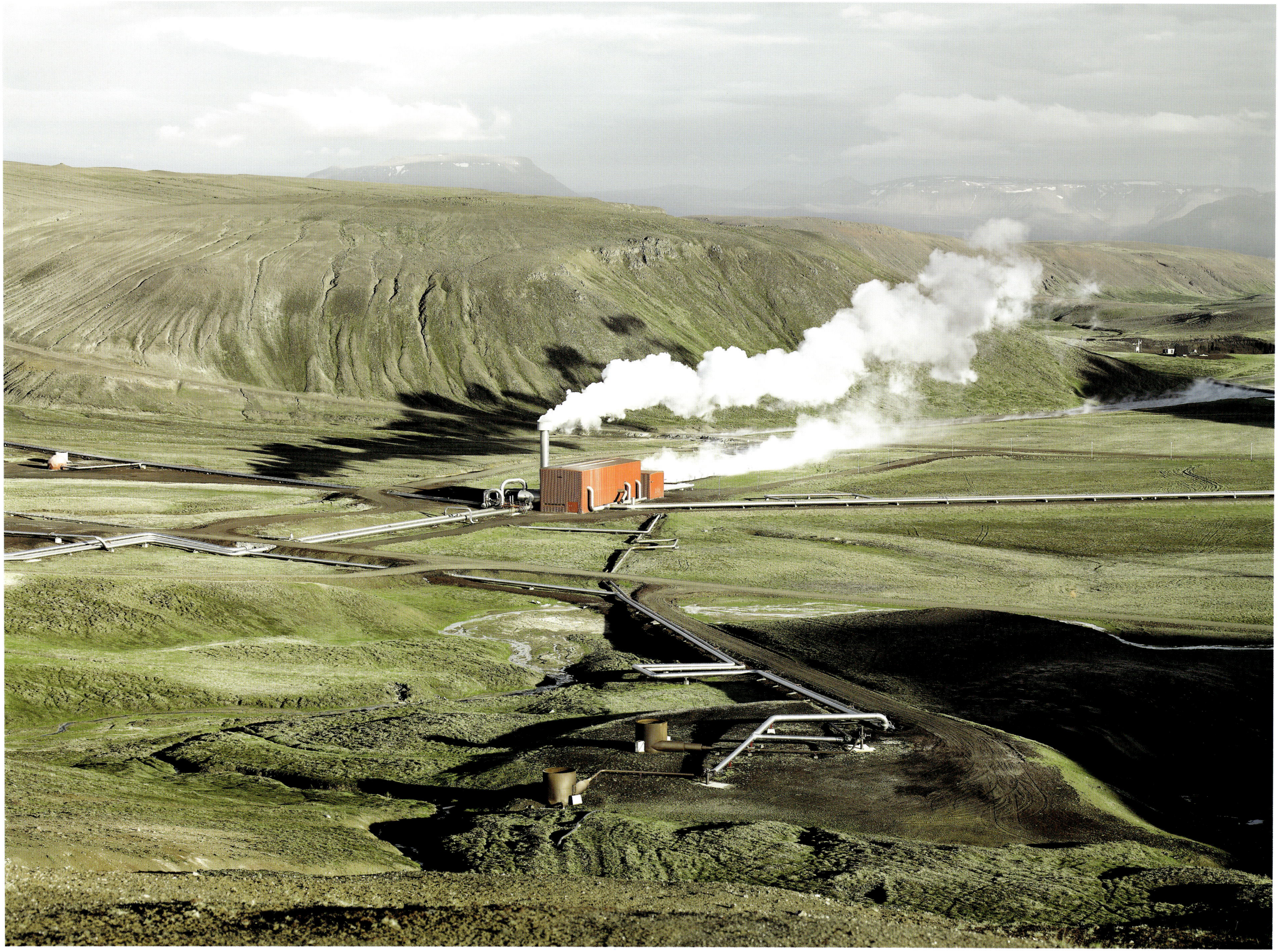

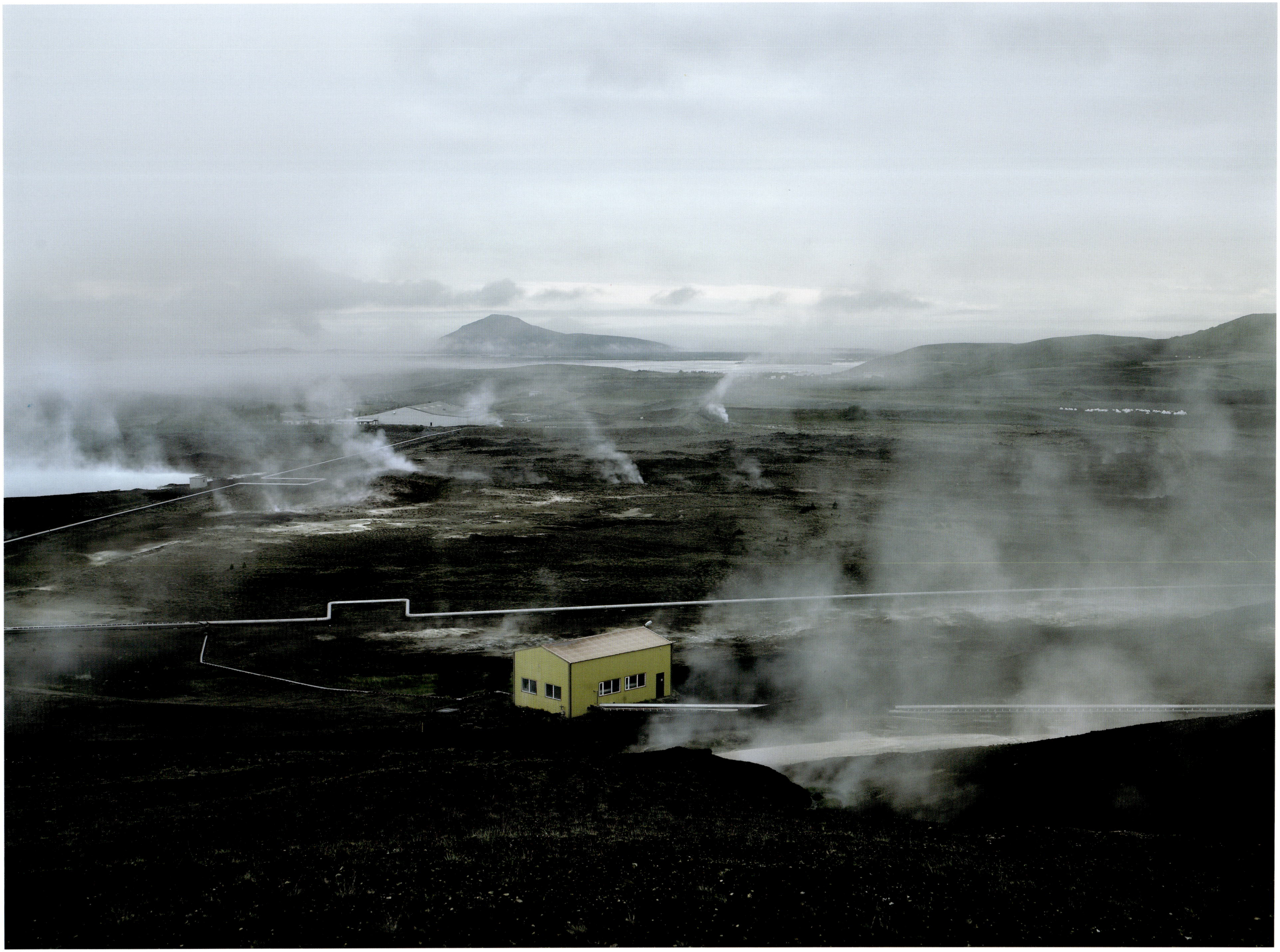

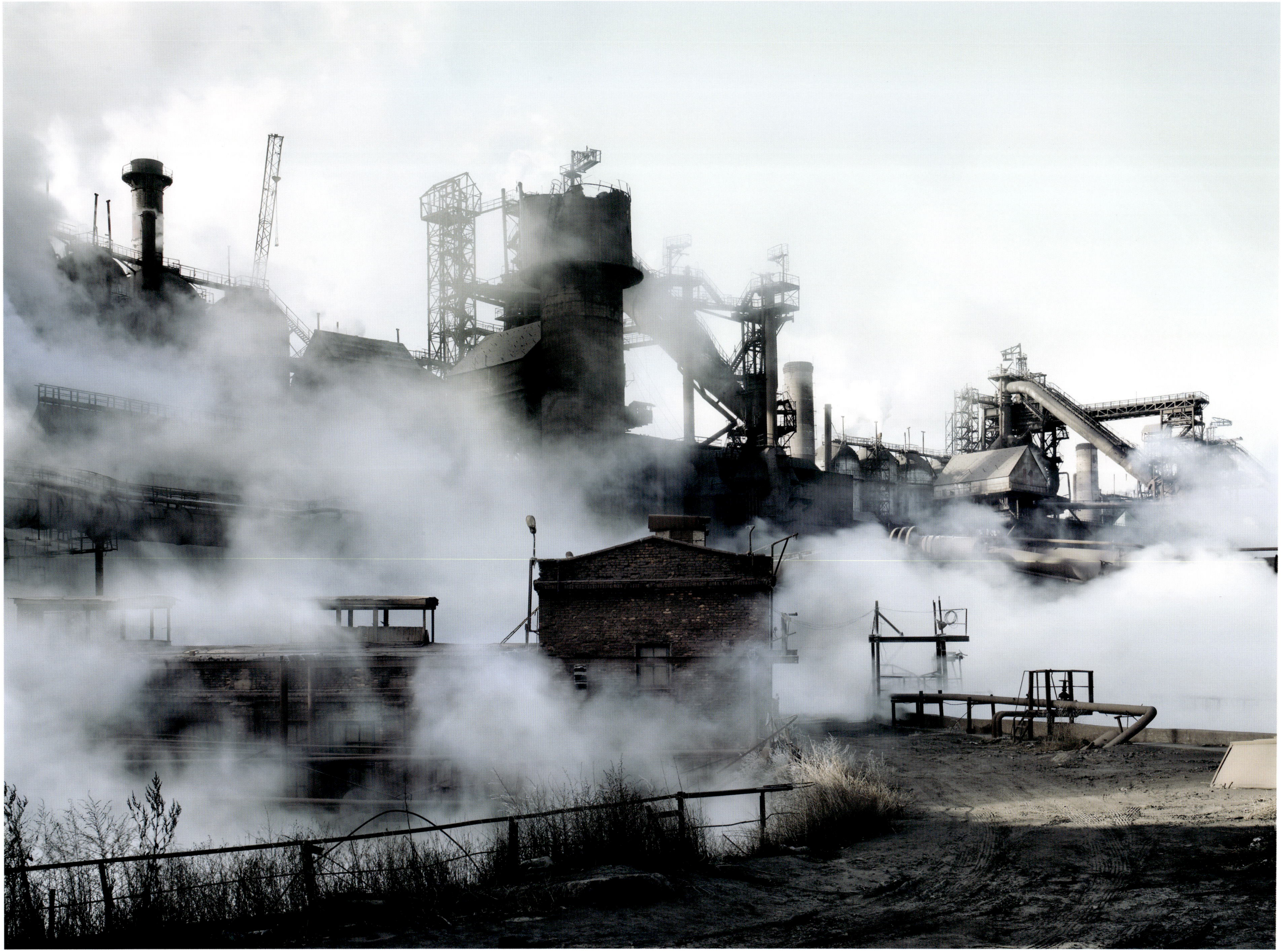

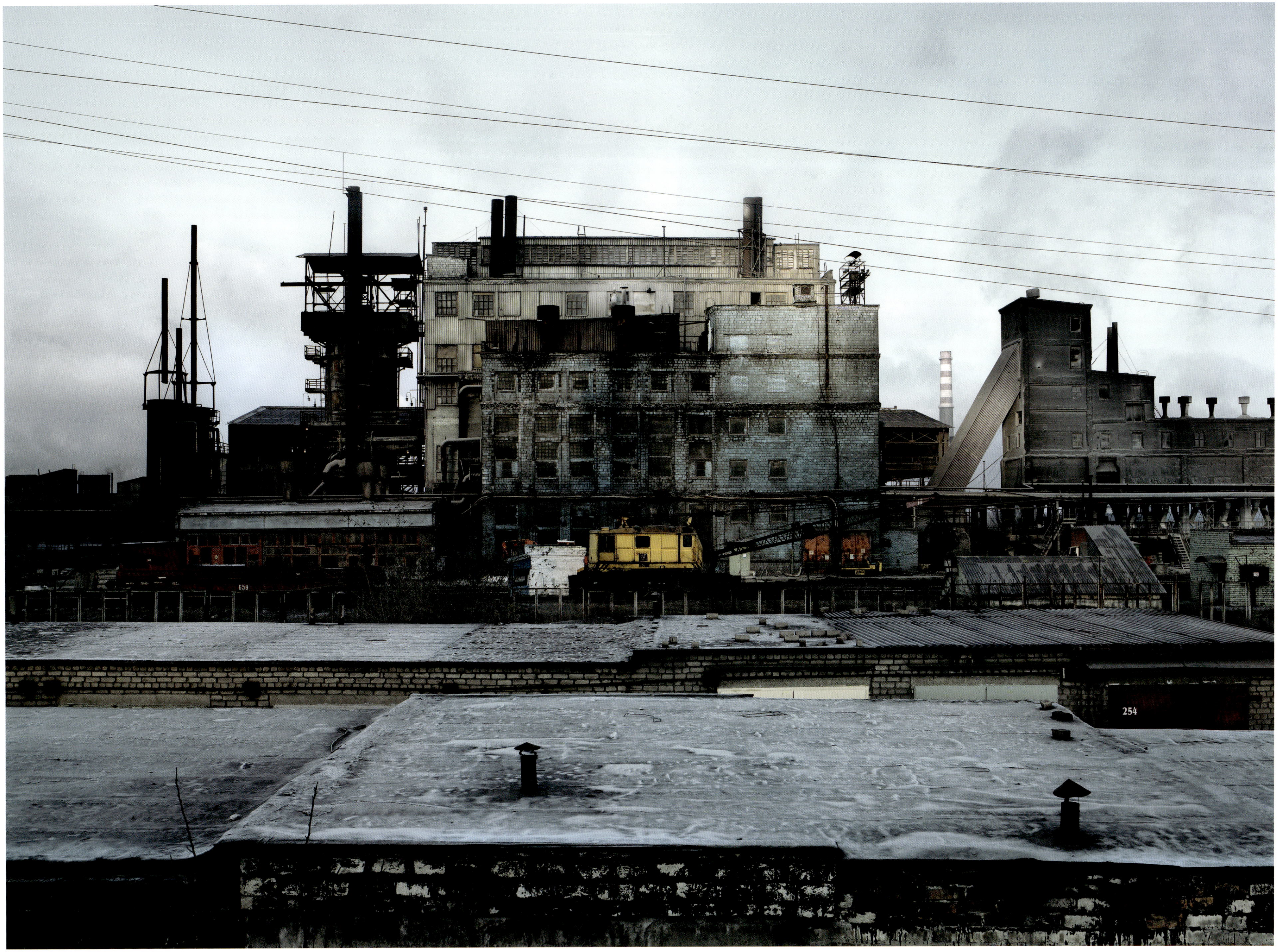
254

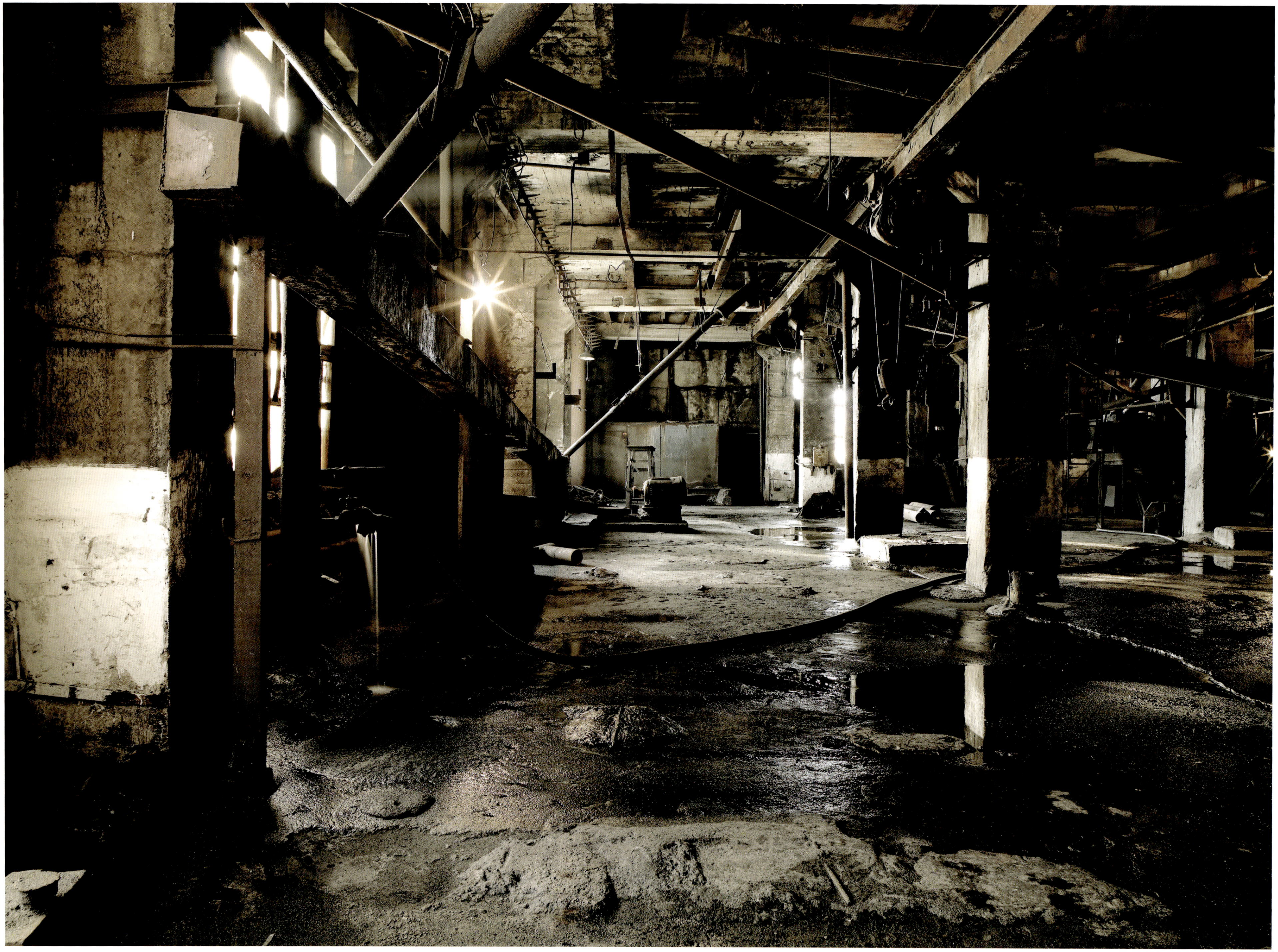

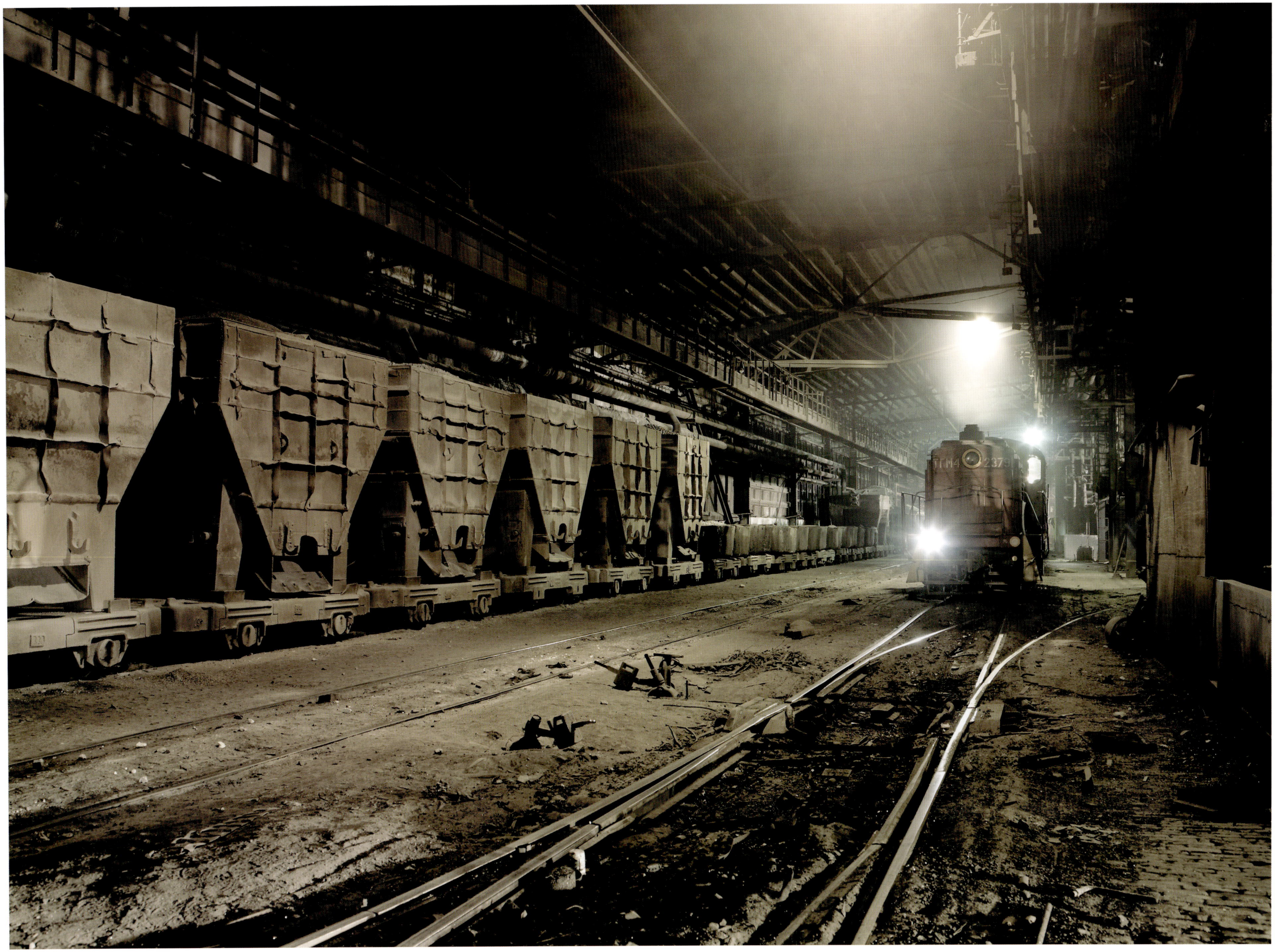

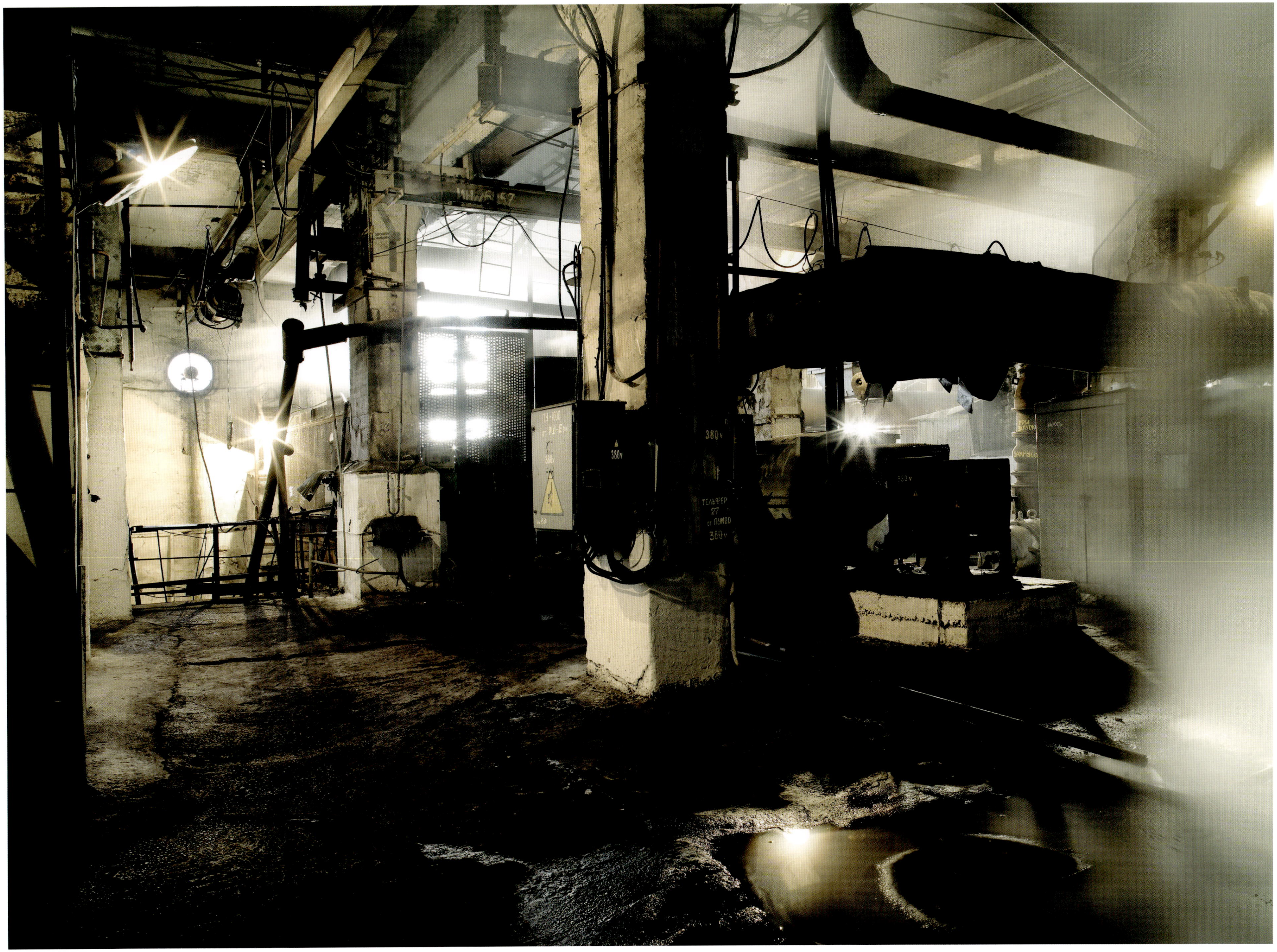
ПУ-1000
380v
380v
380v
ТЕЛЬФЕР
27
380v

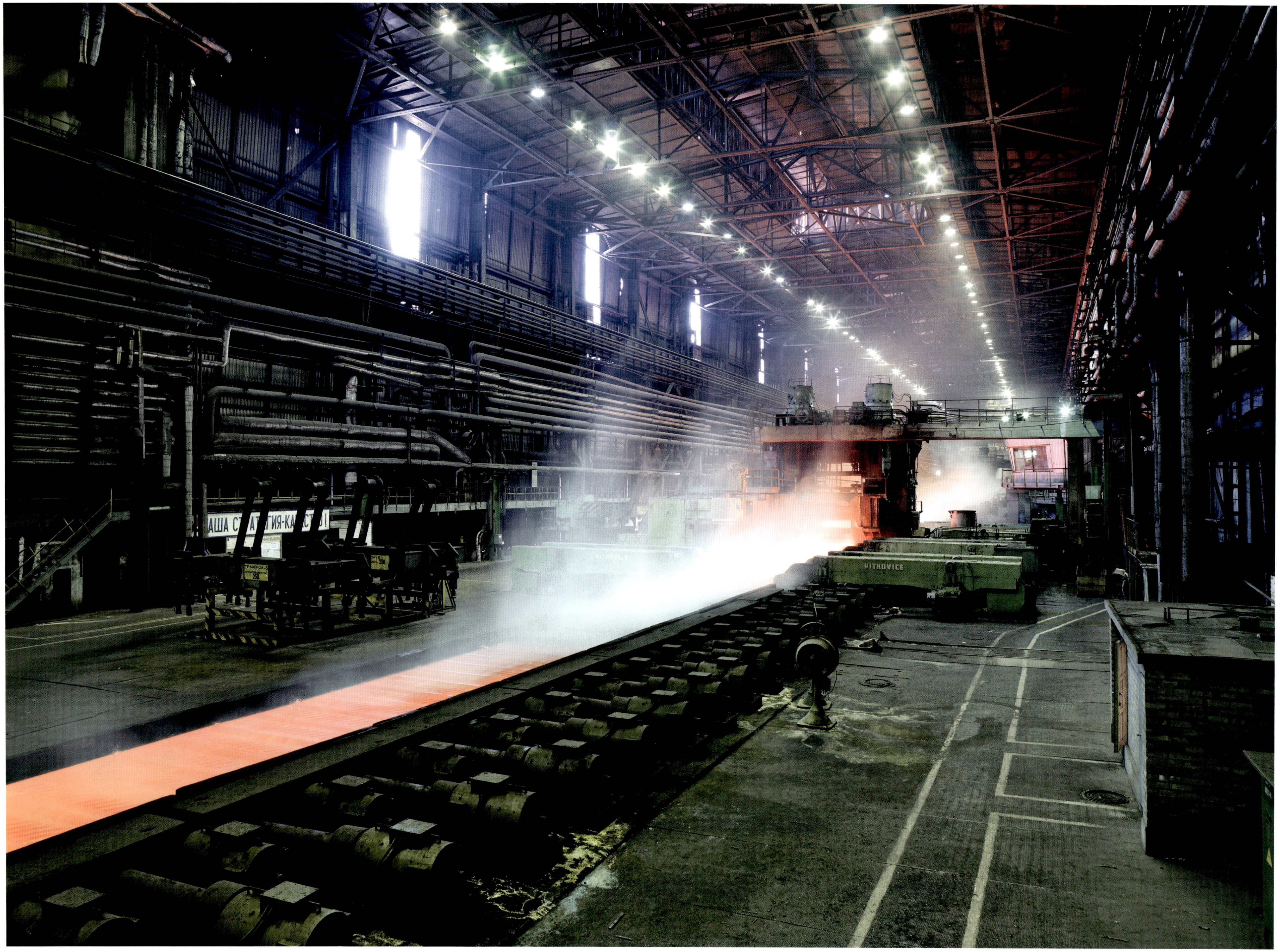
VITKOVICE

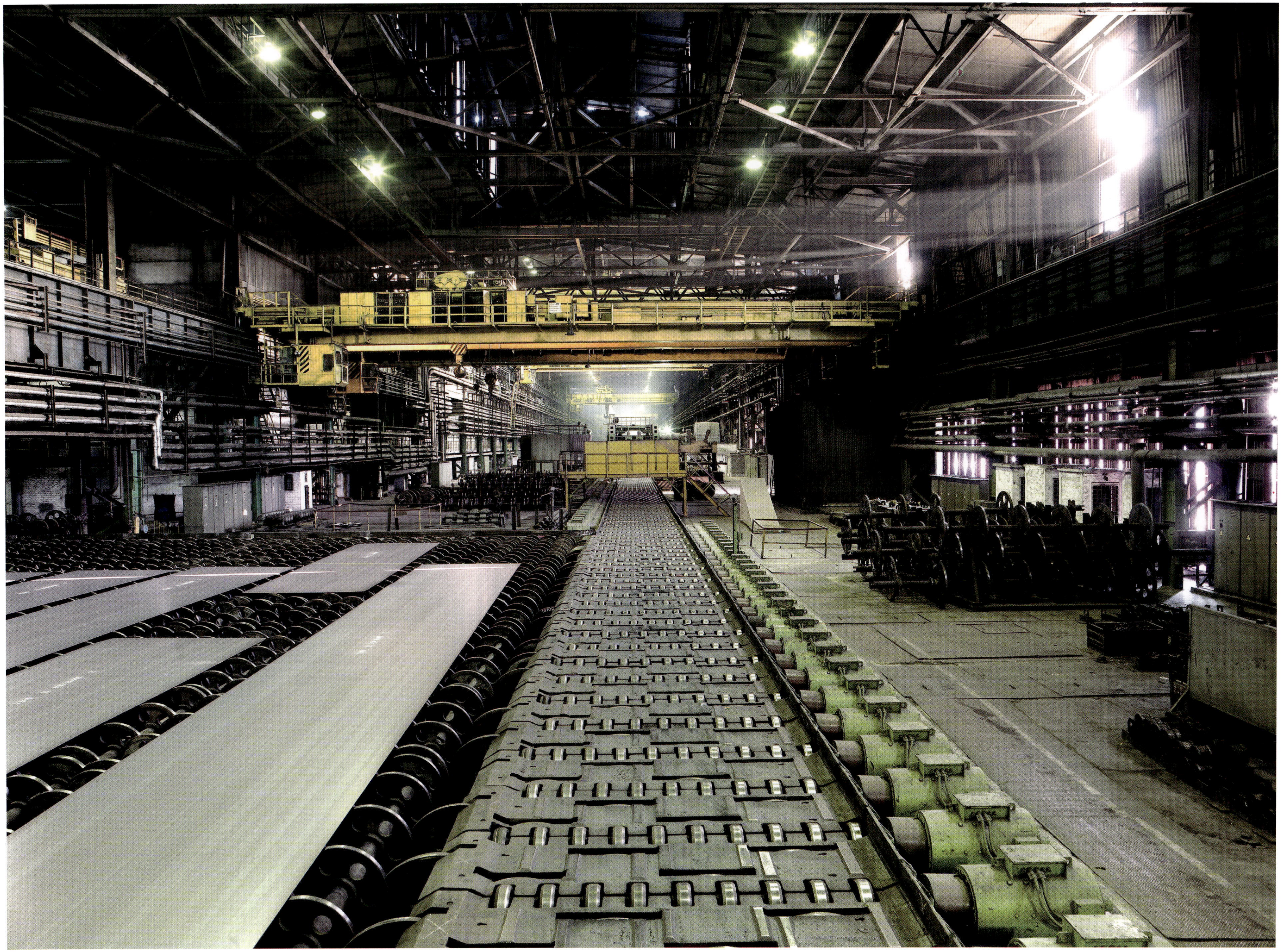

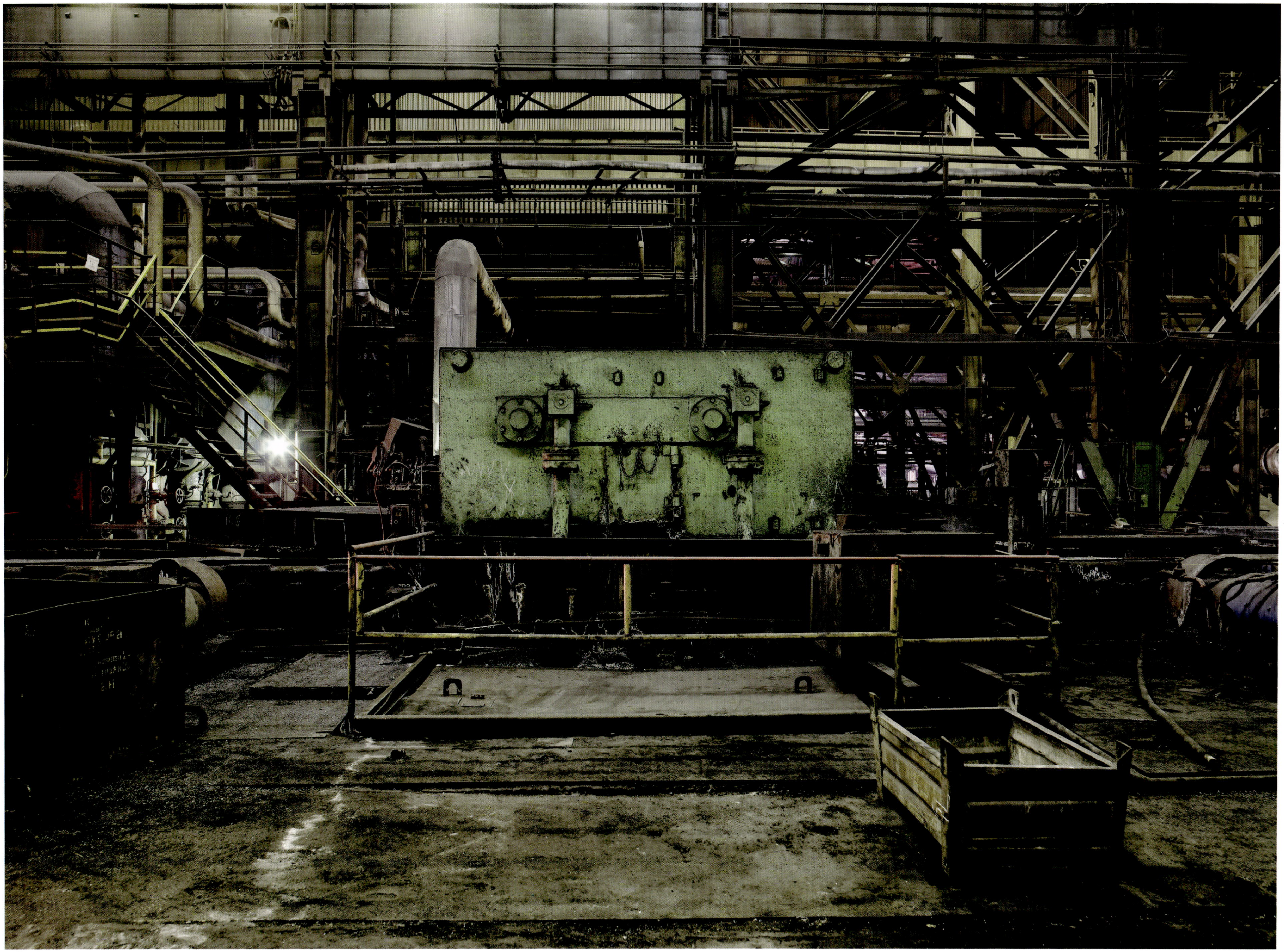

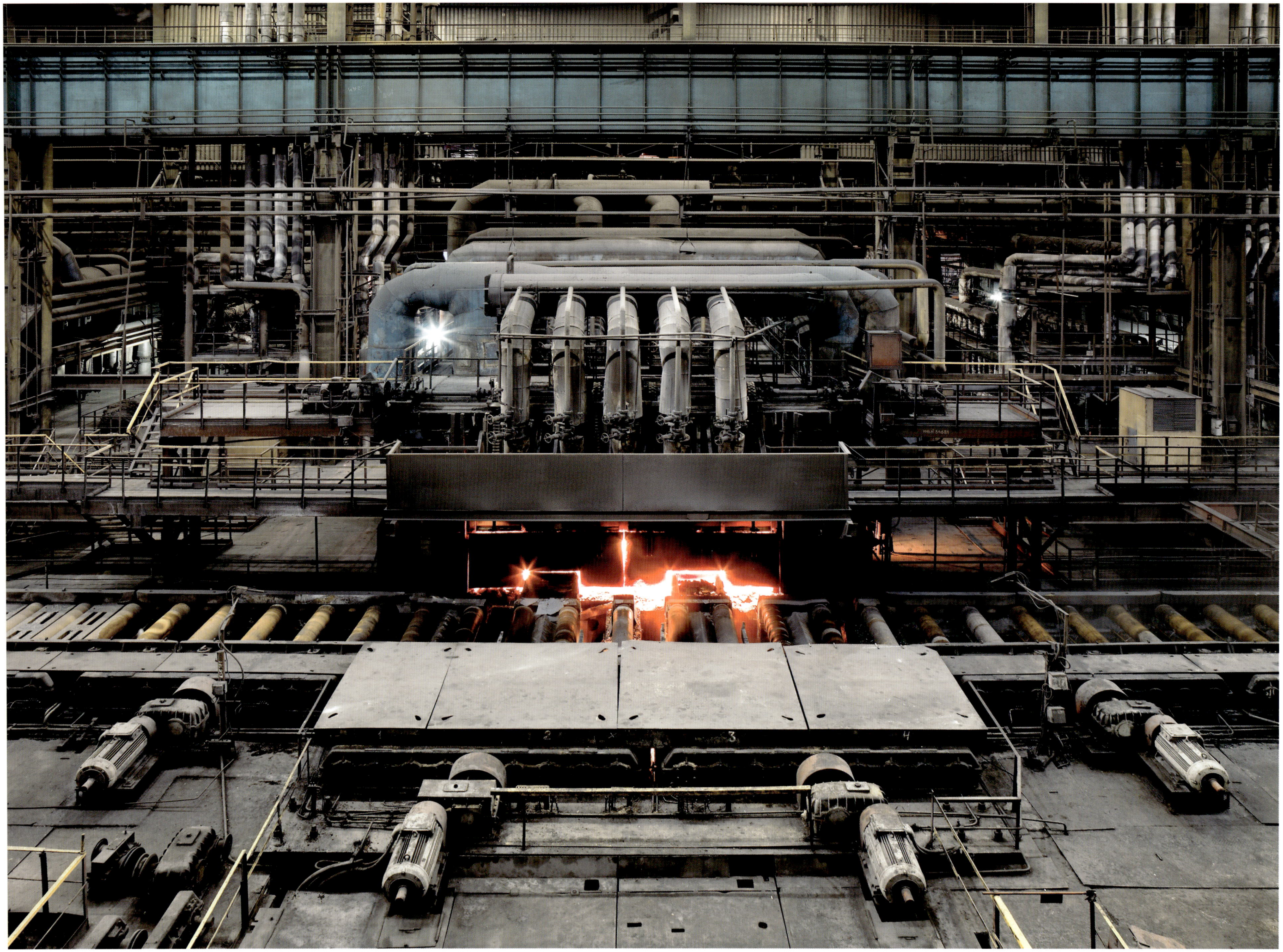

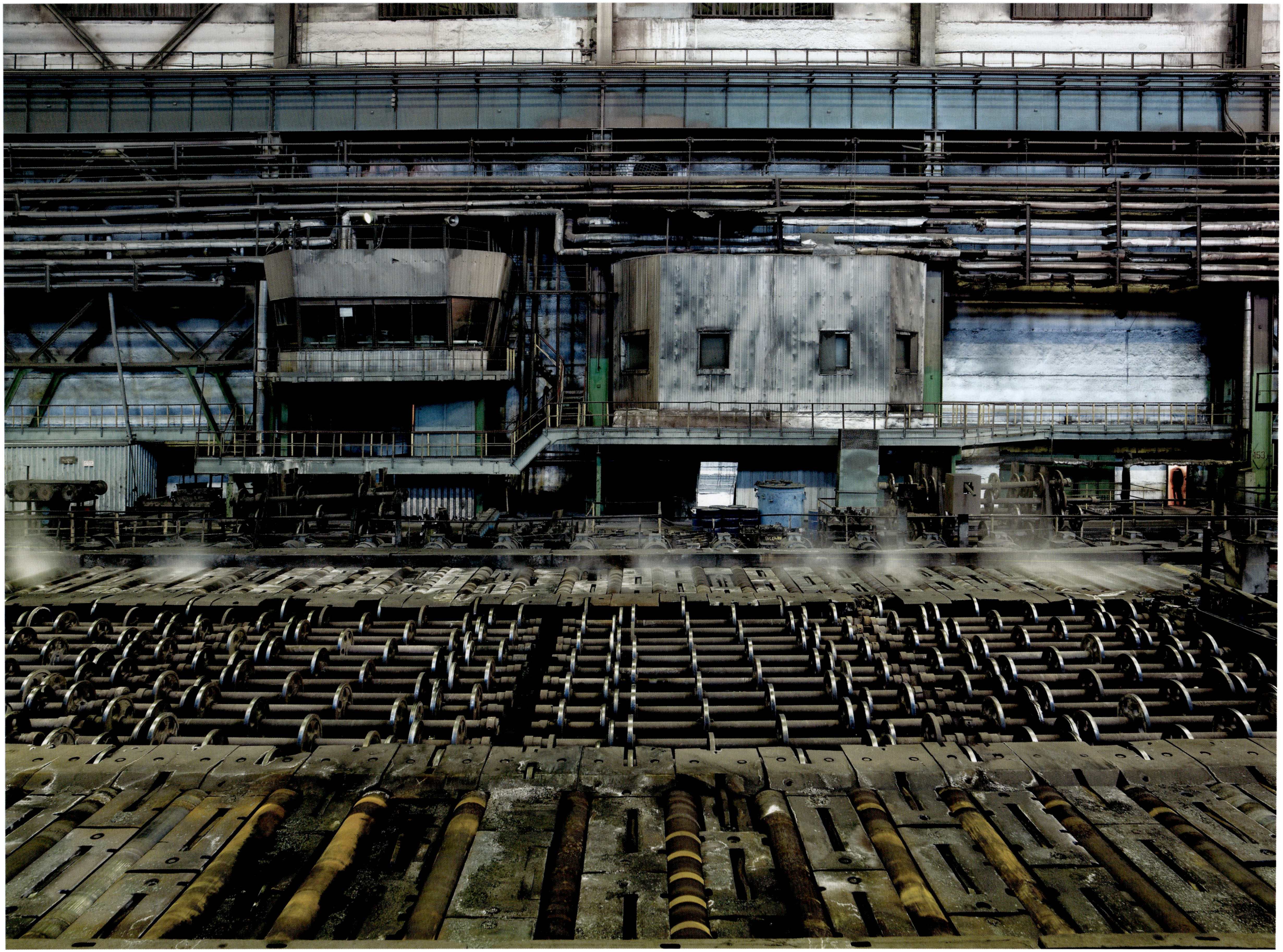

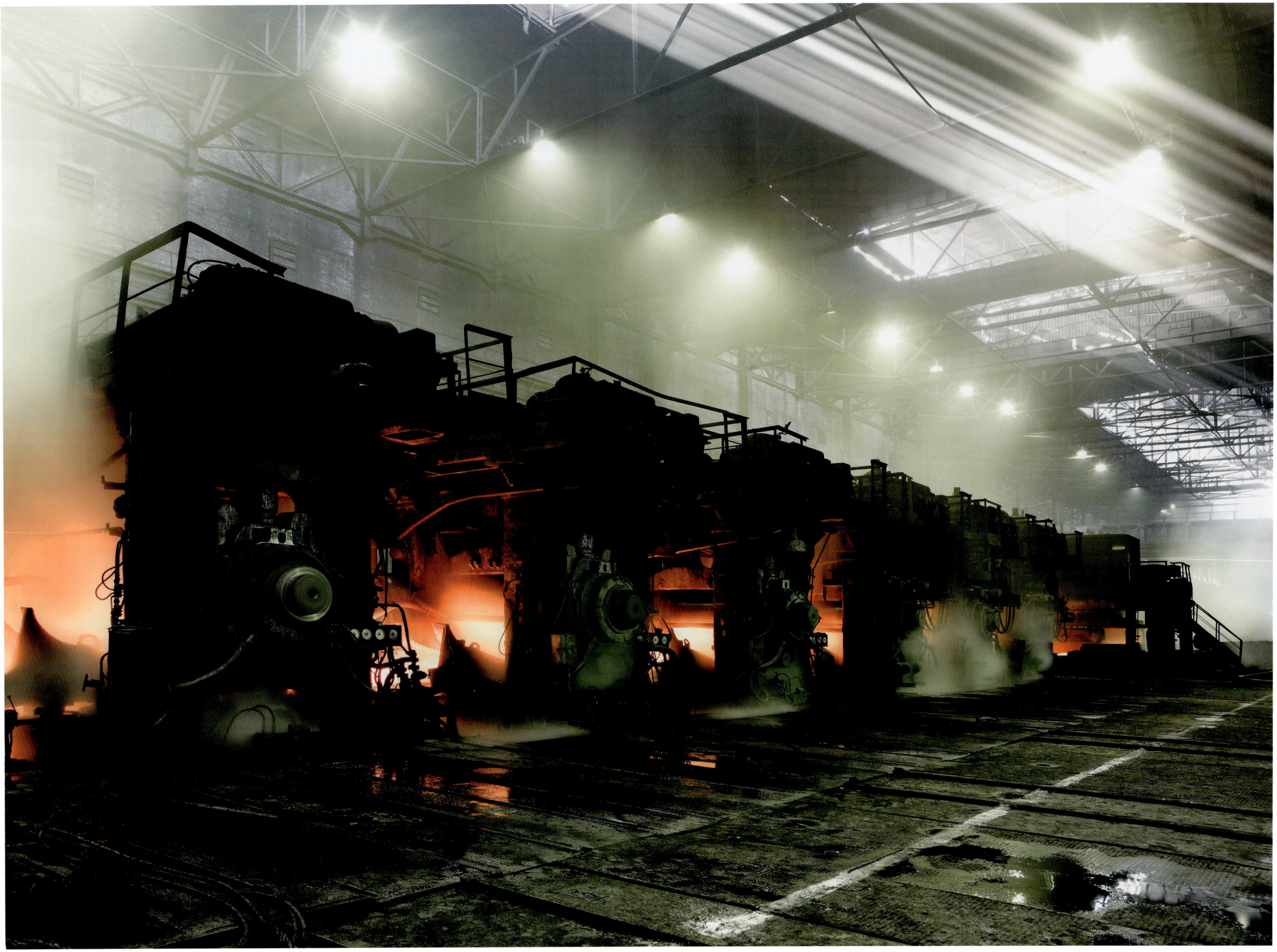

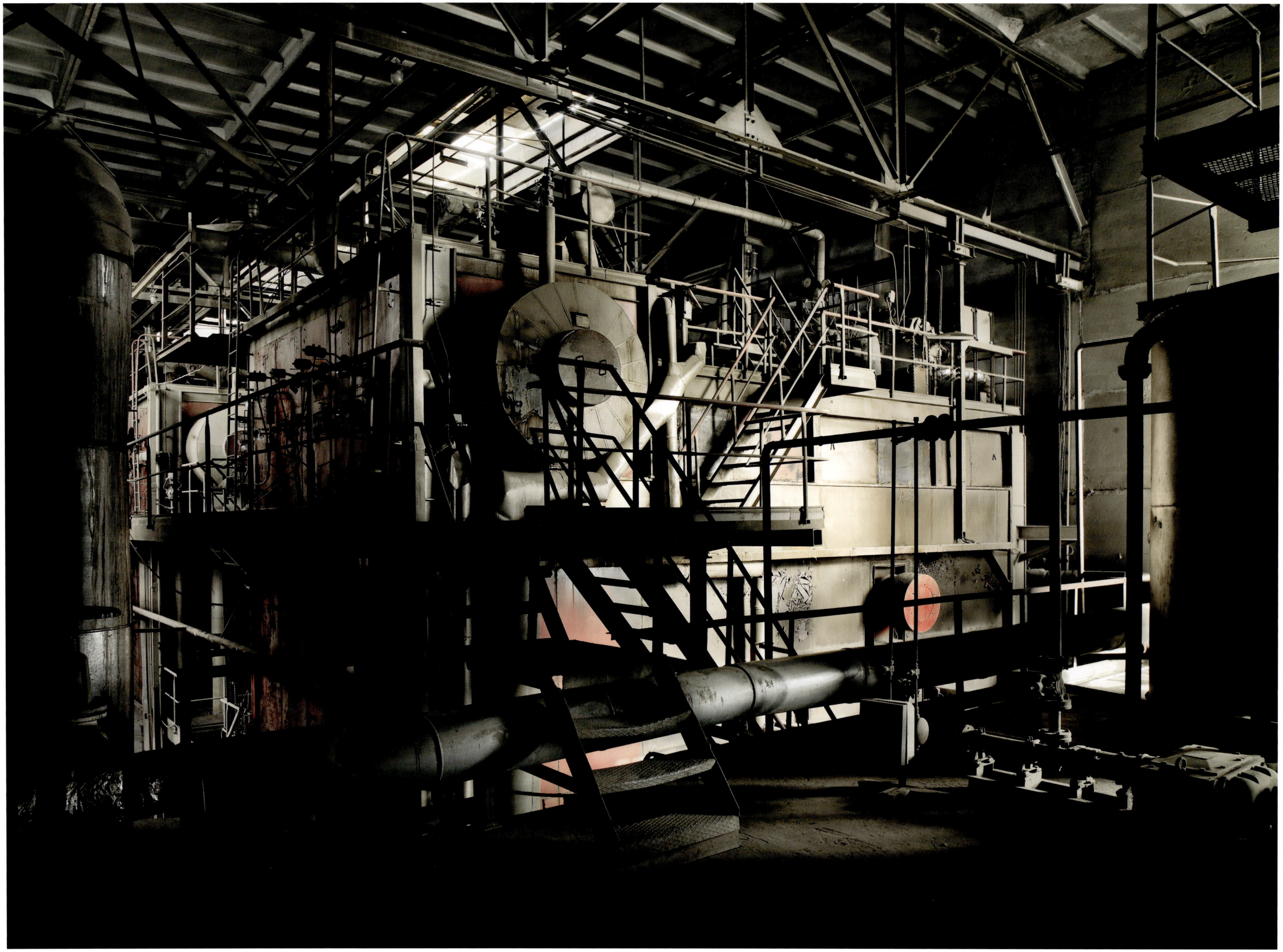

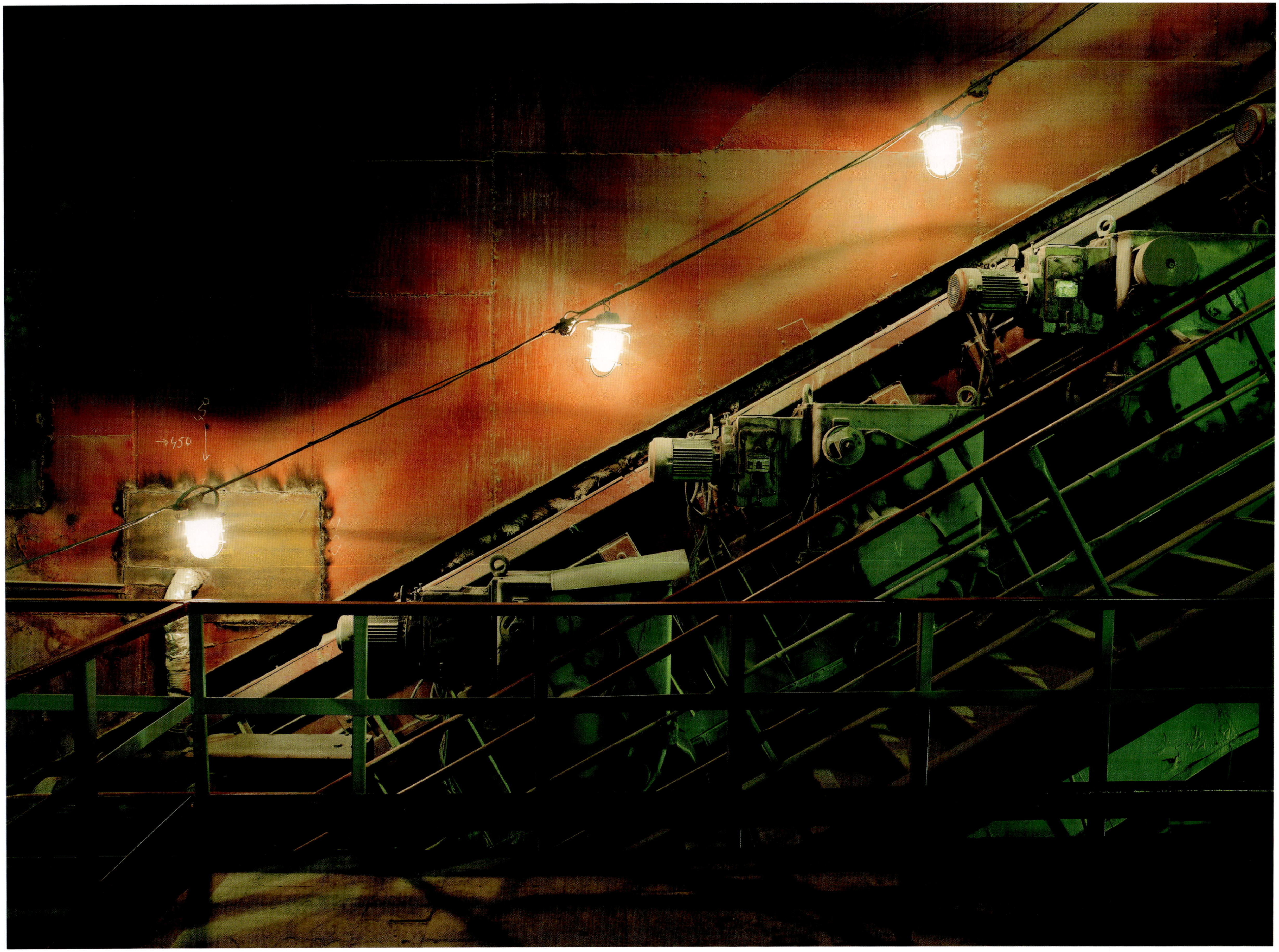

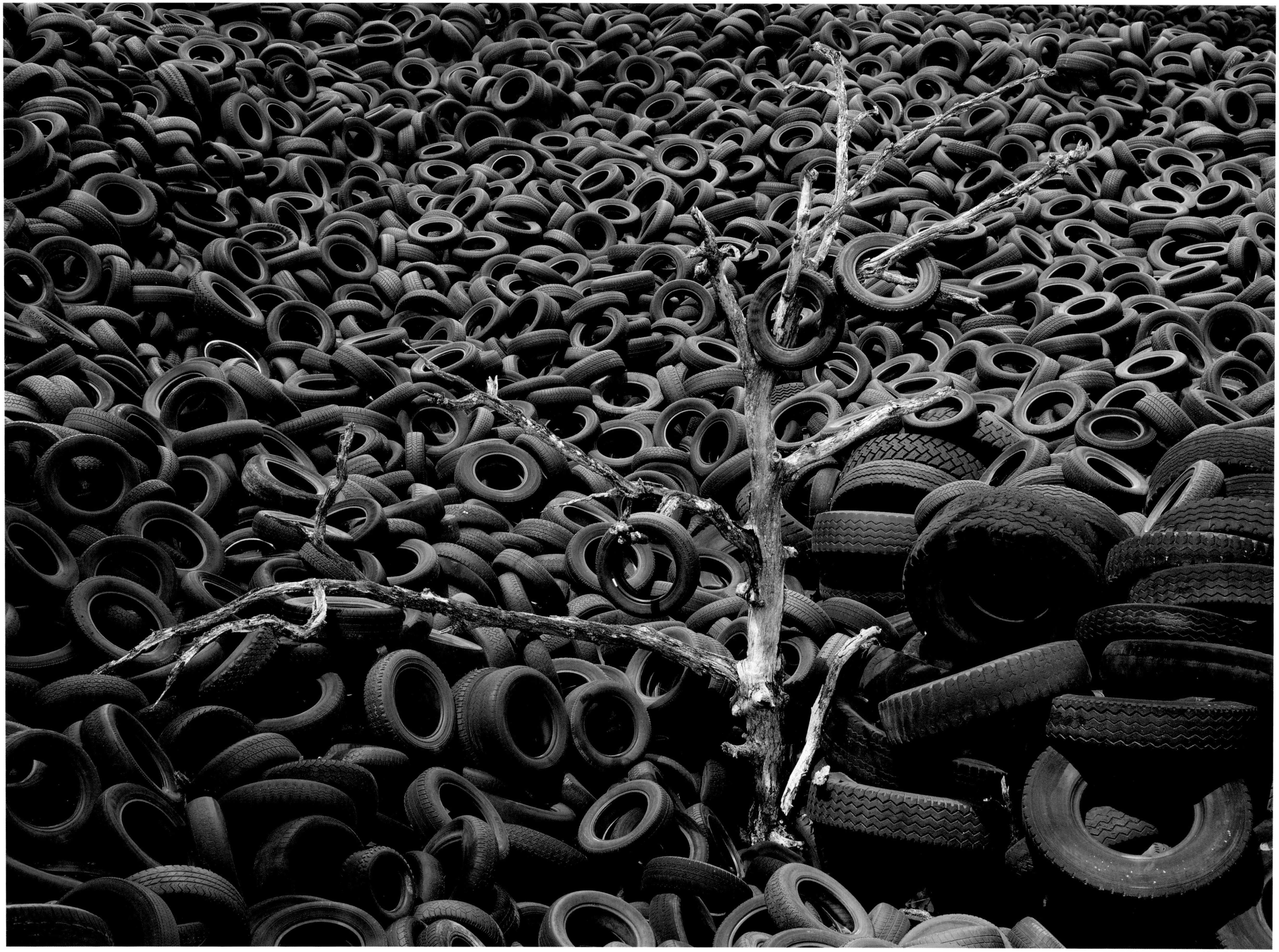

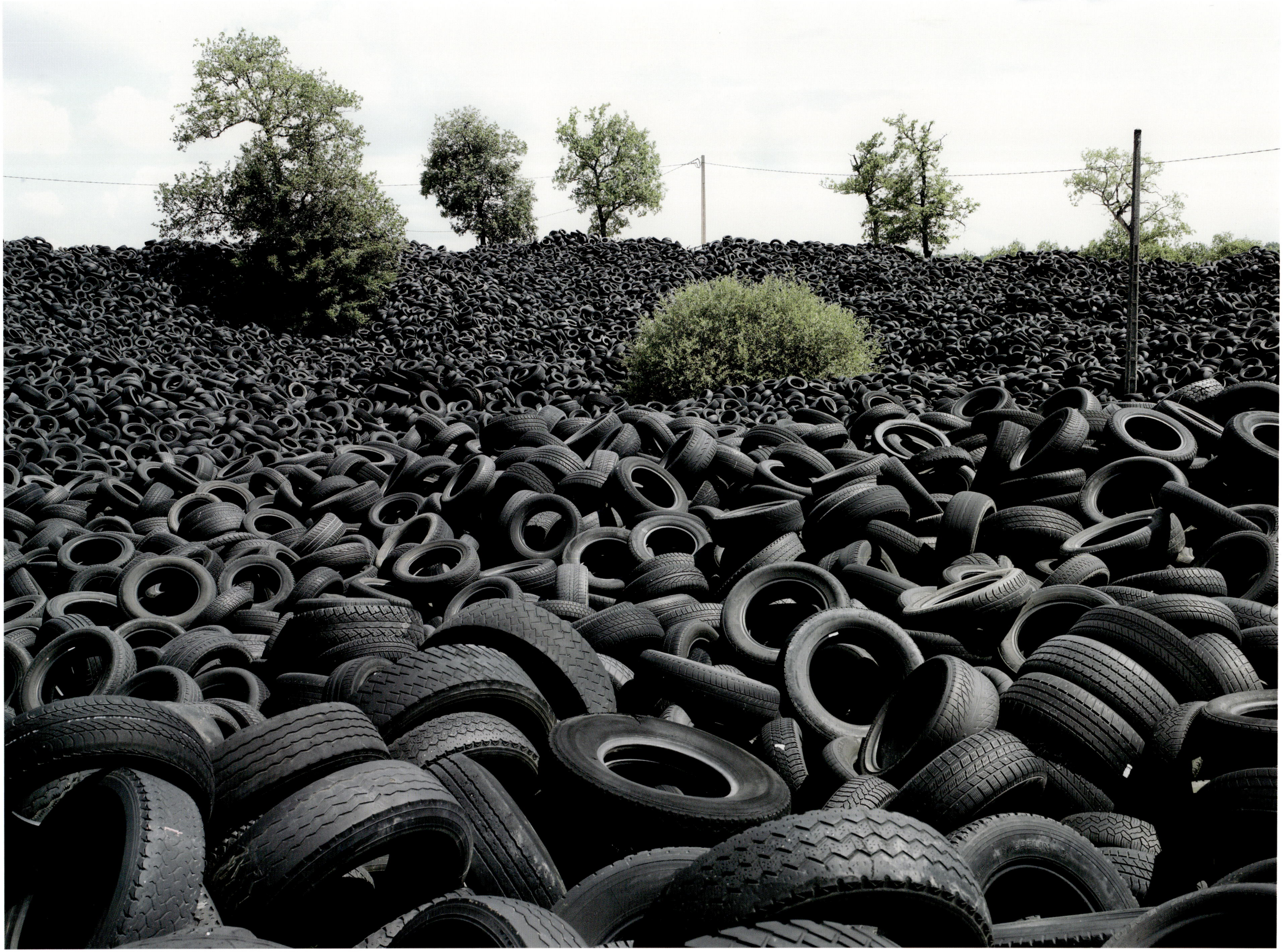

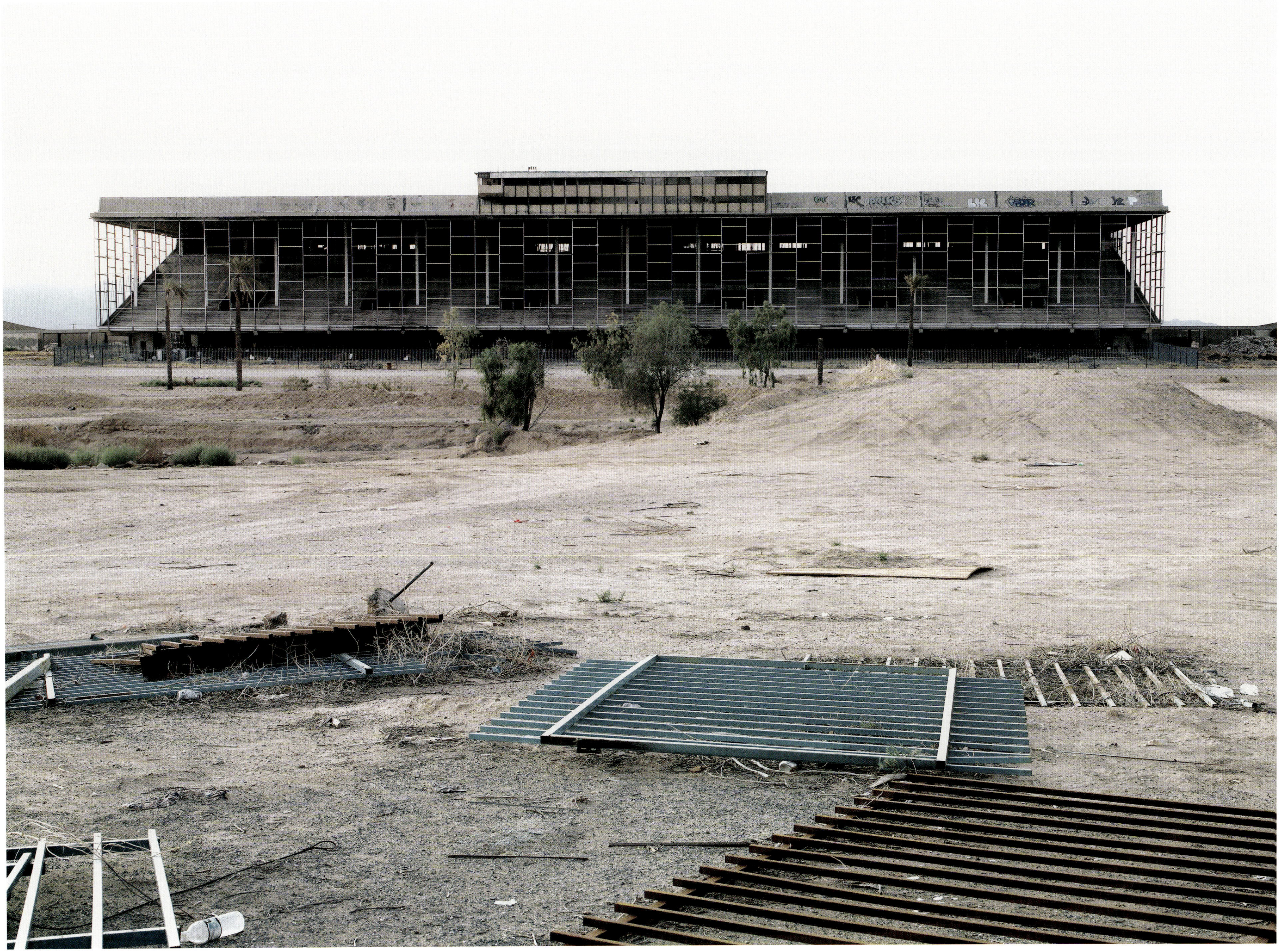

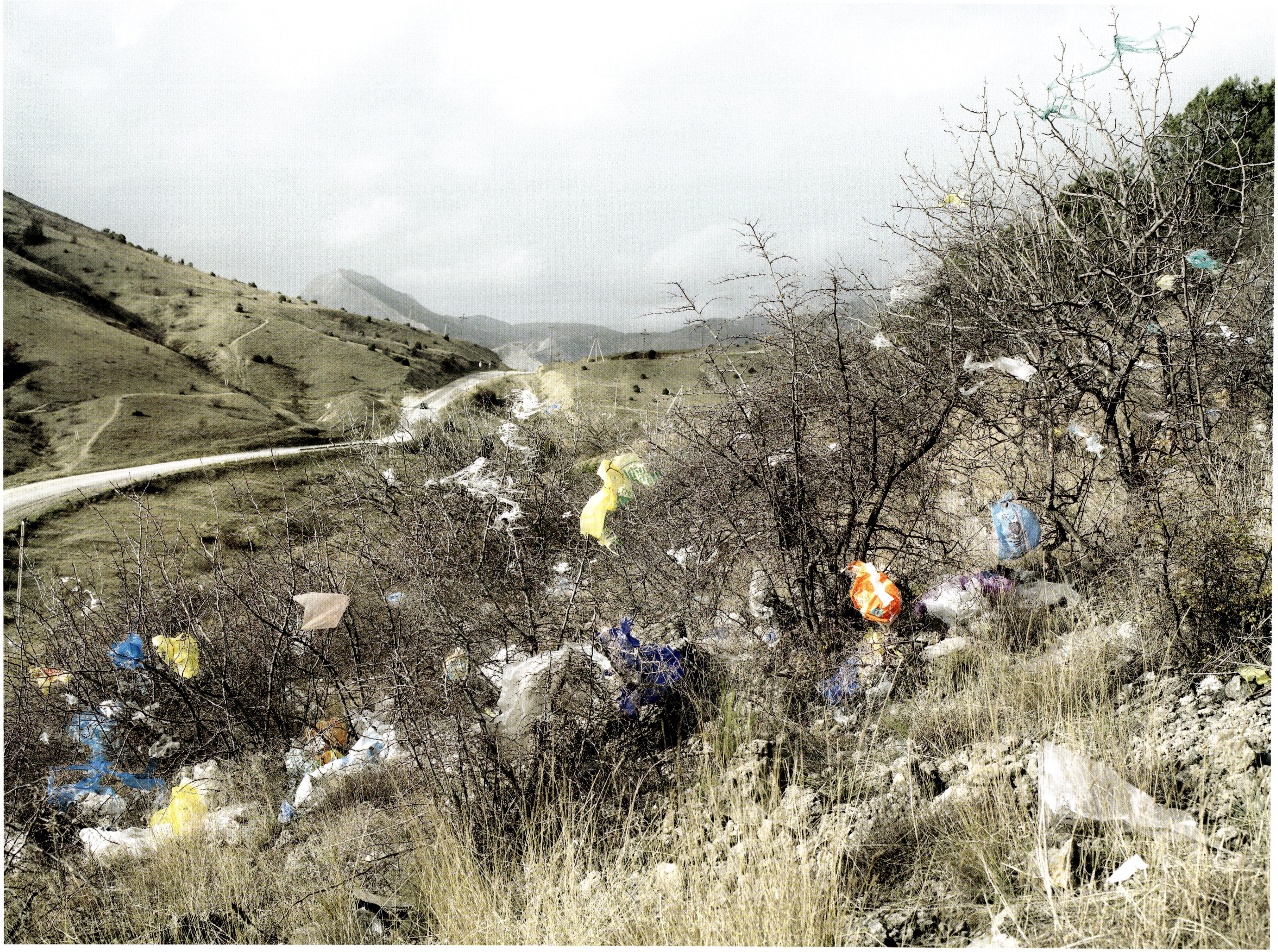

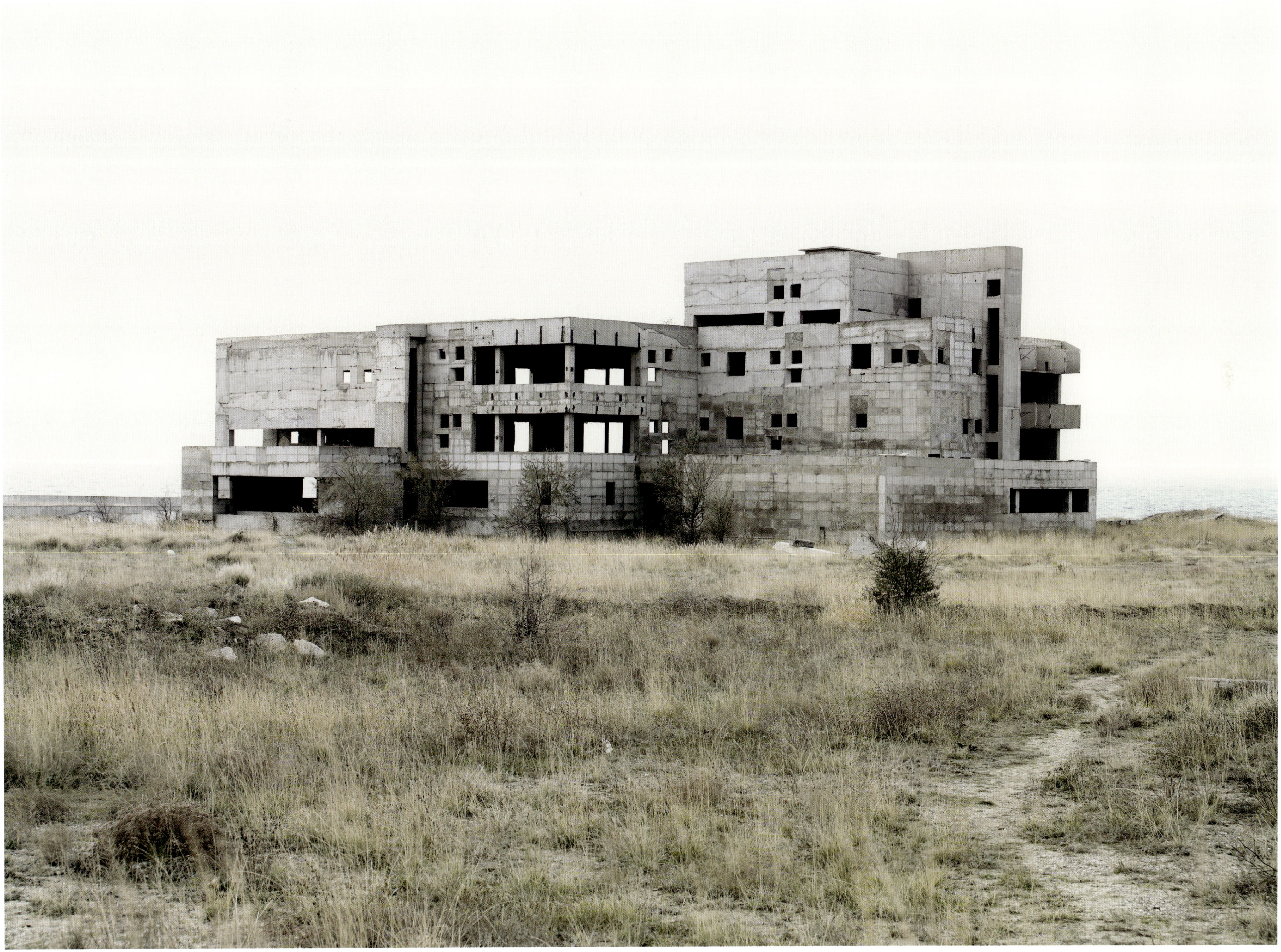

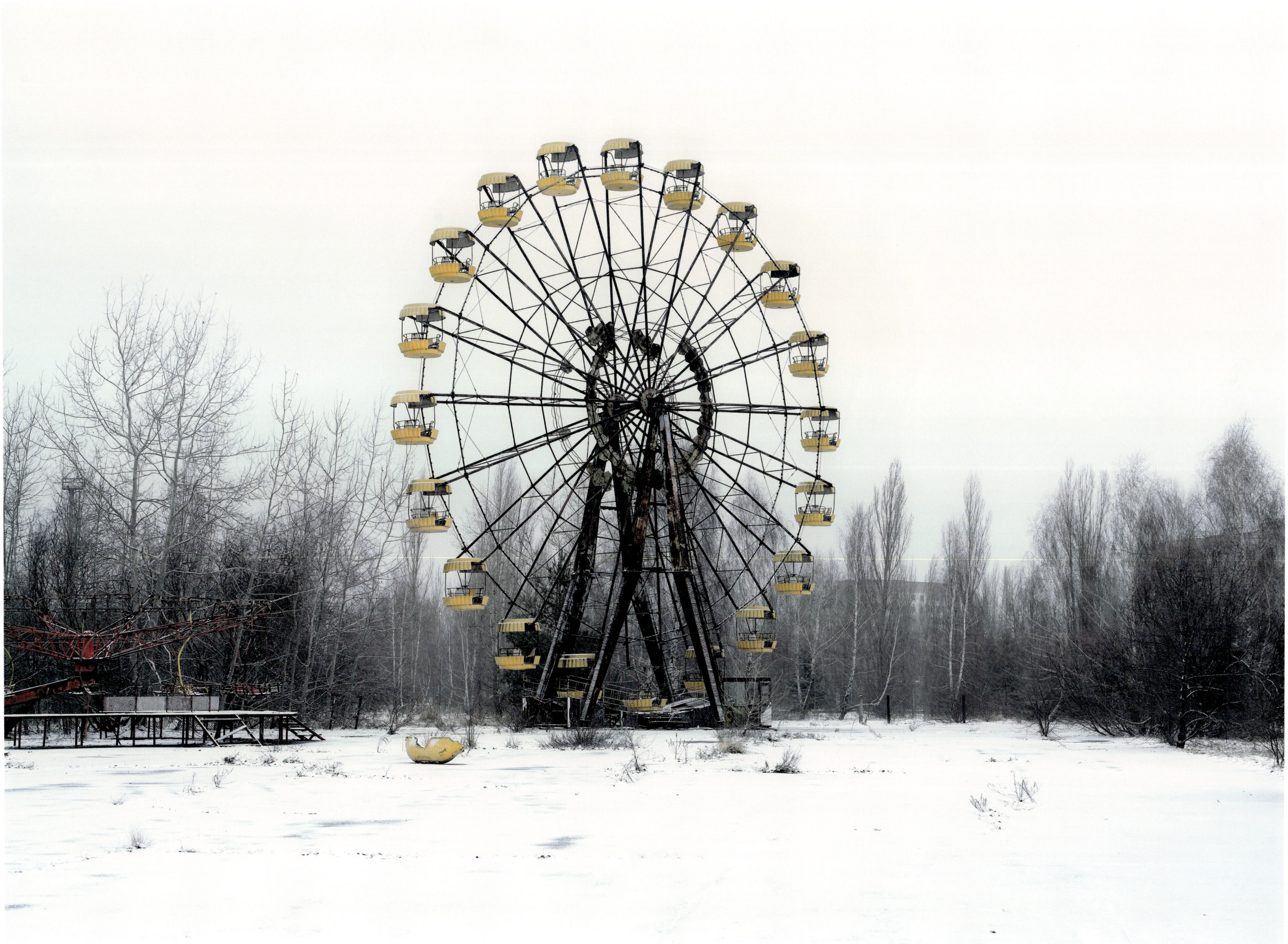

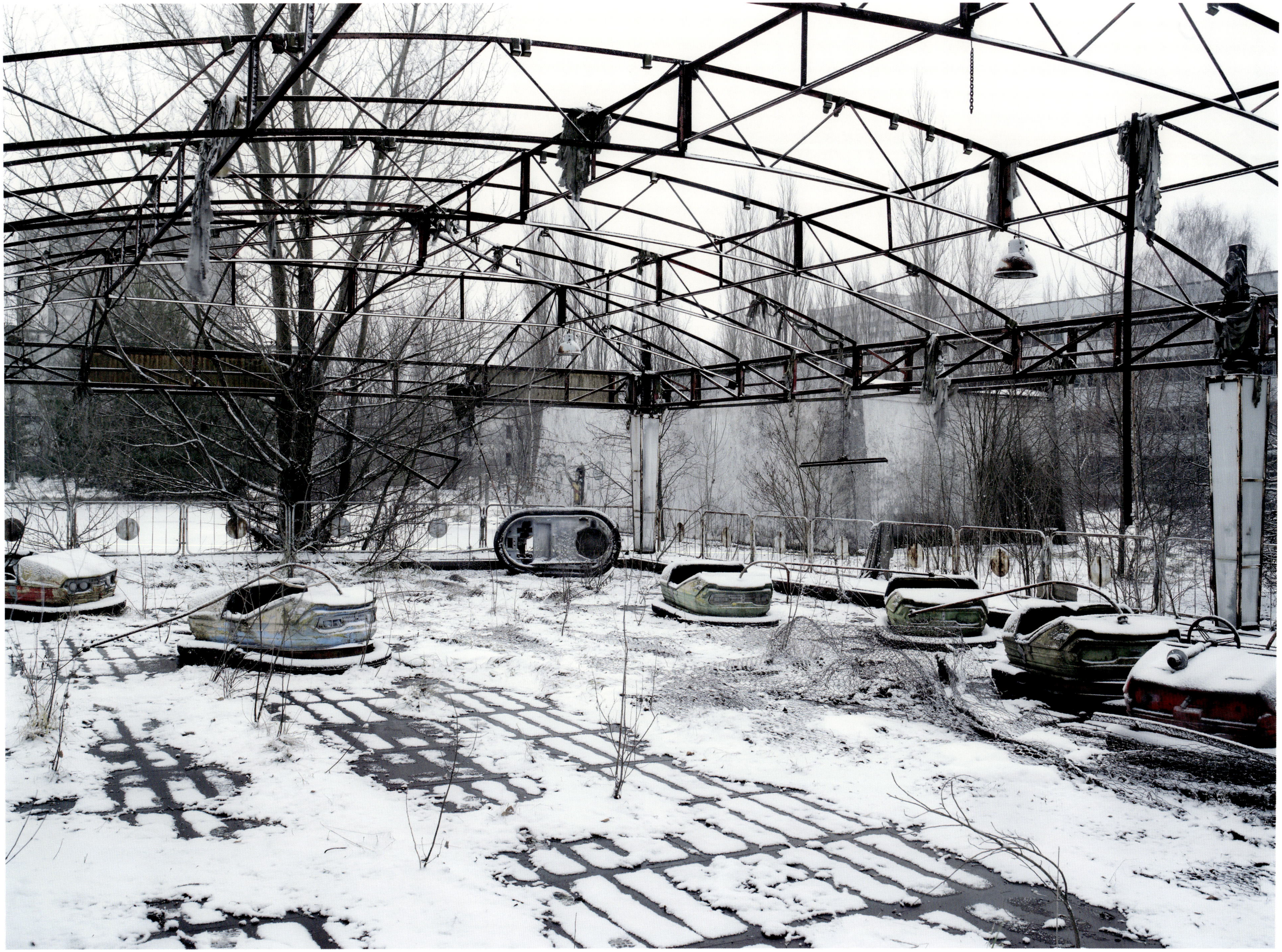

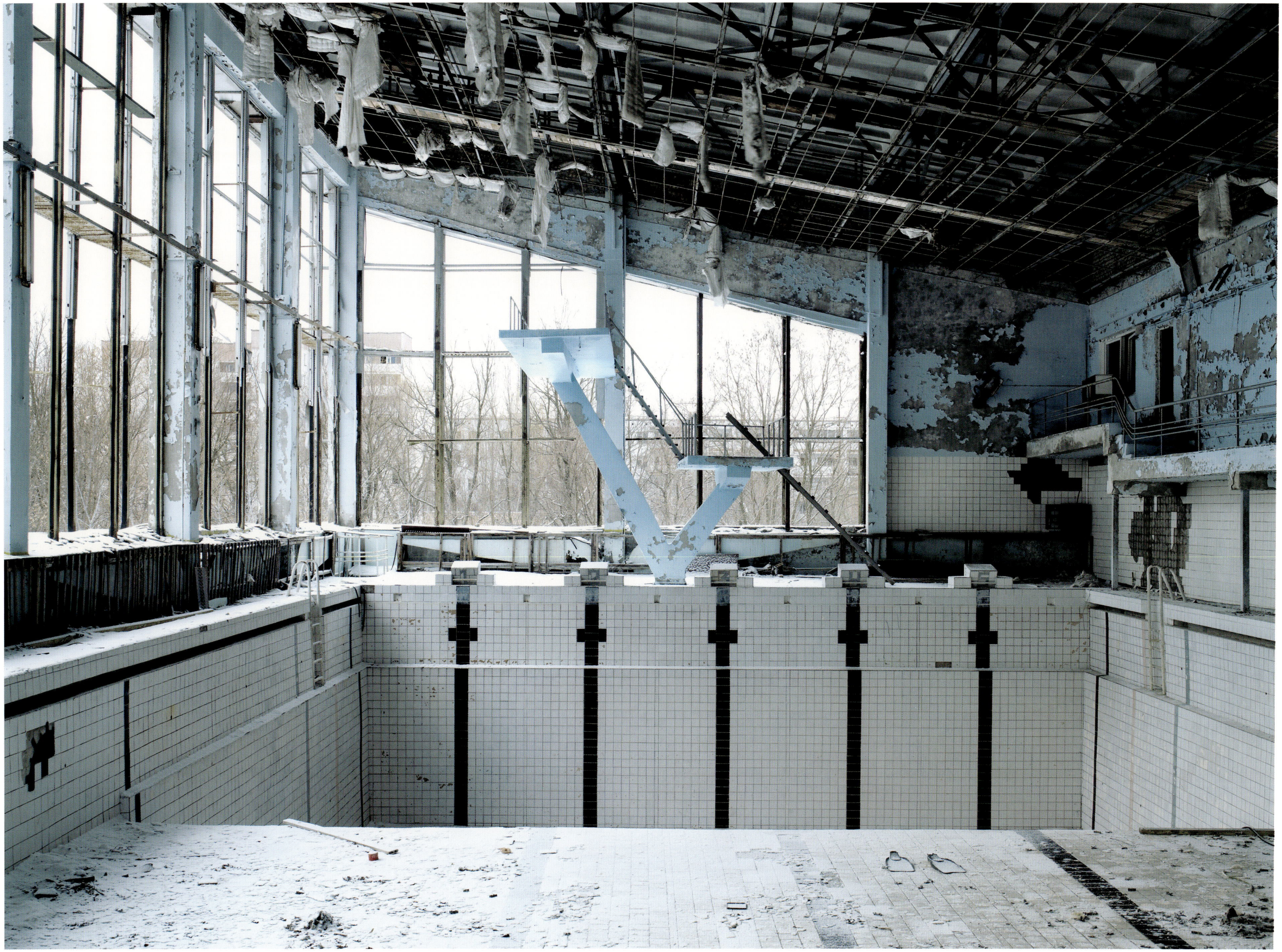

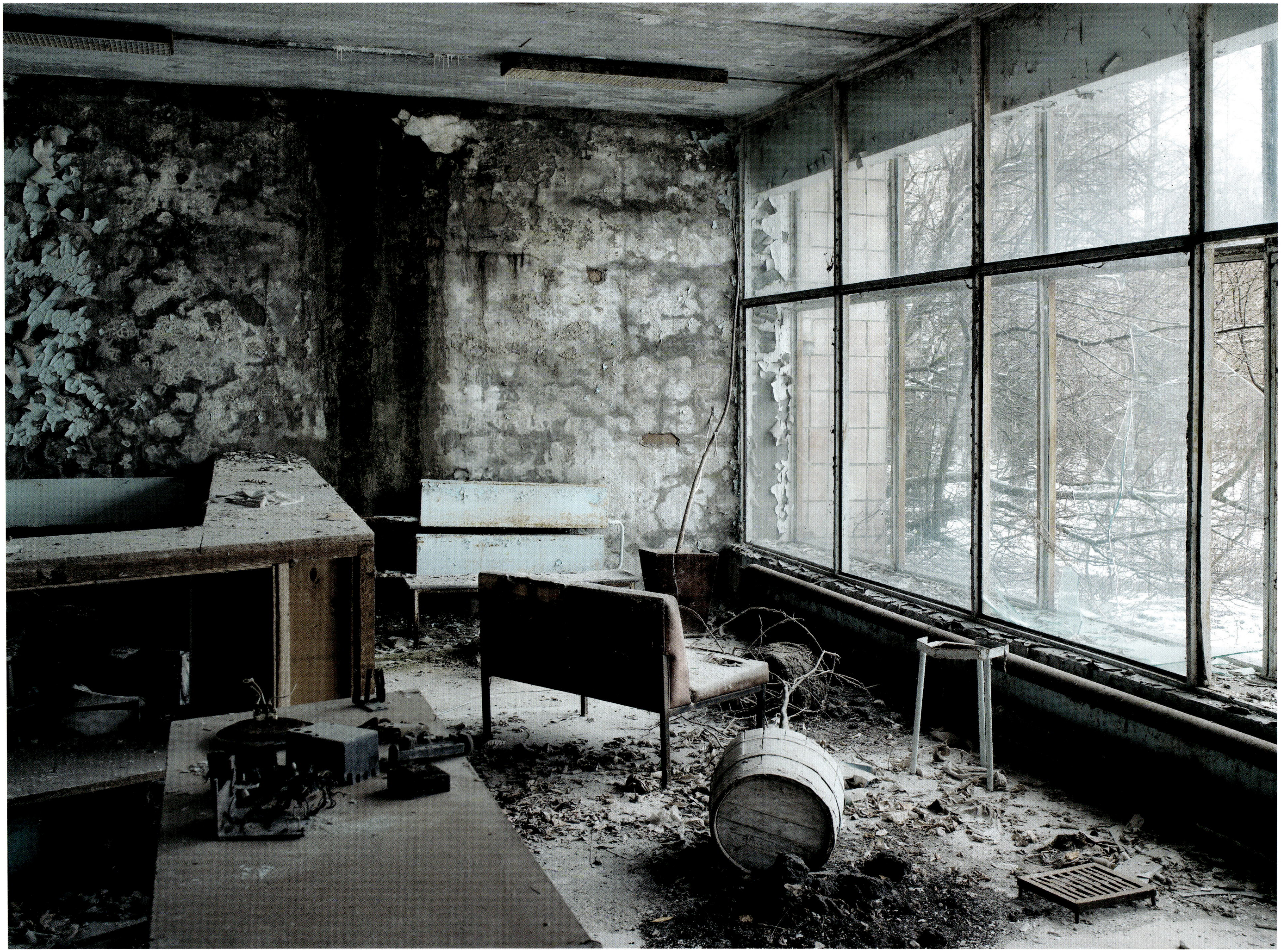

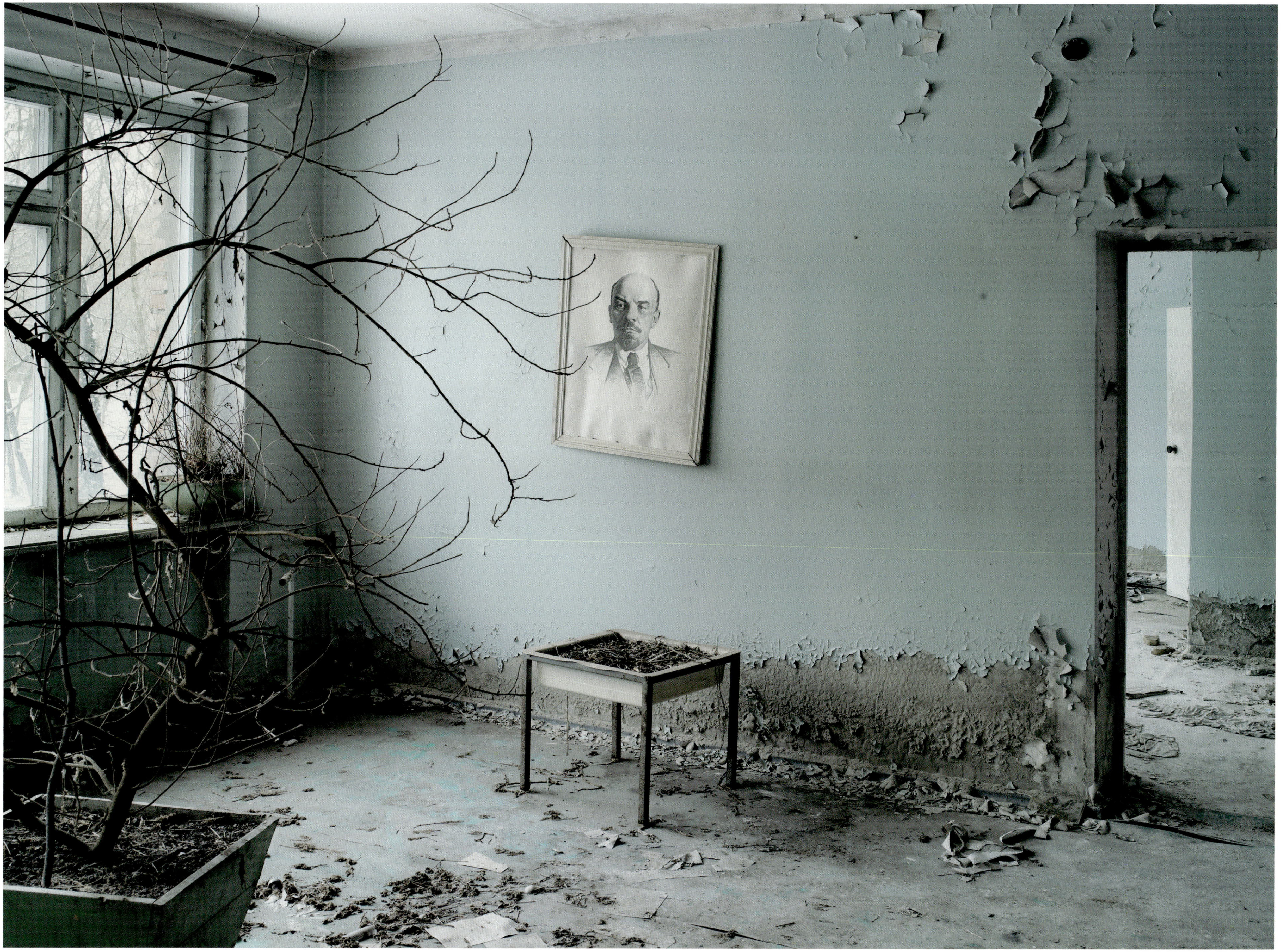

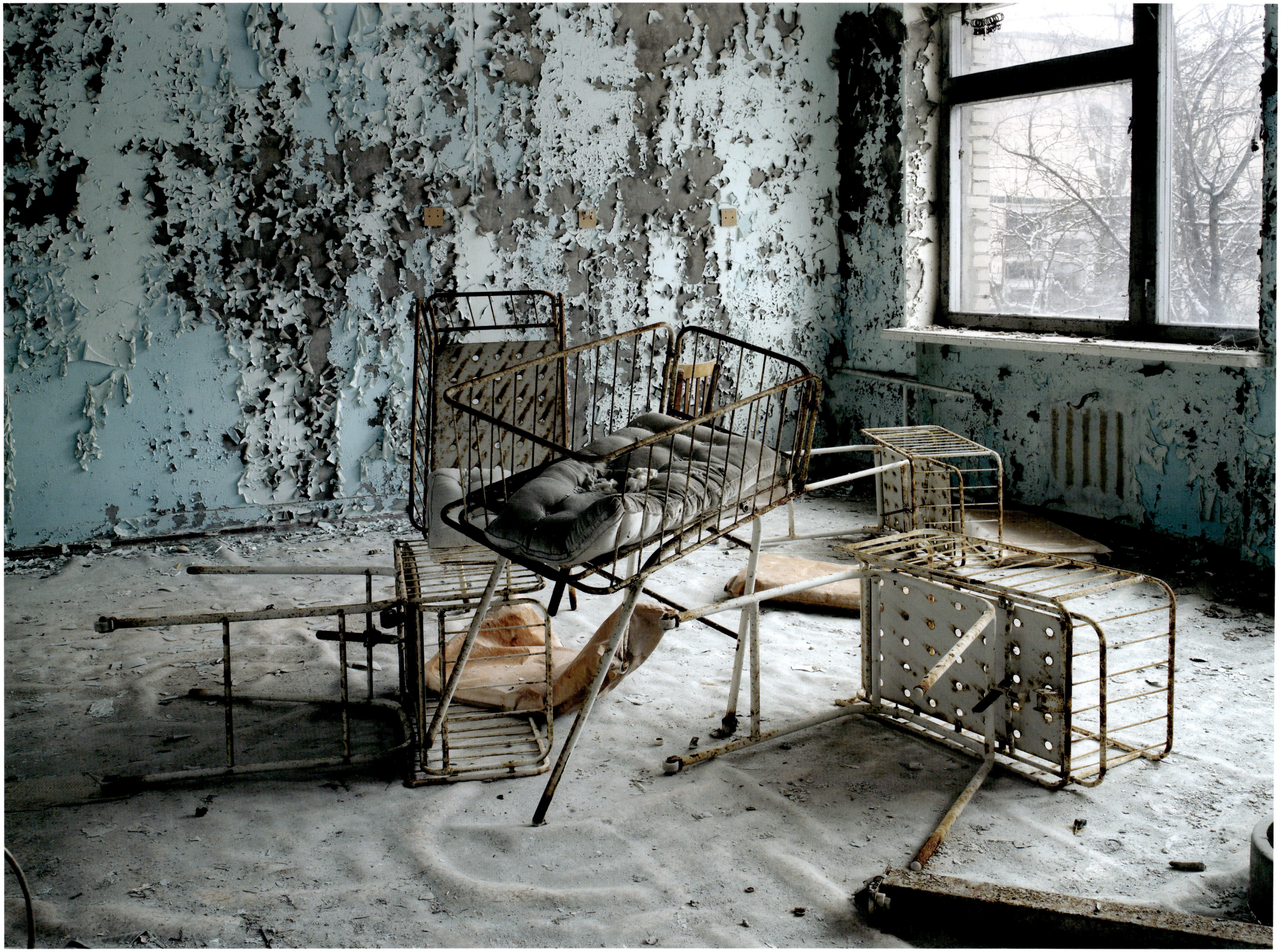

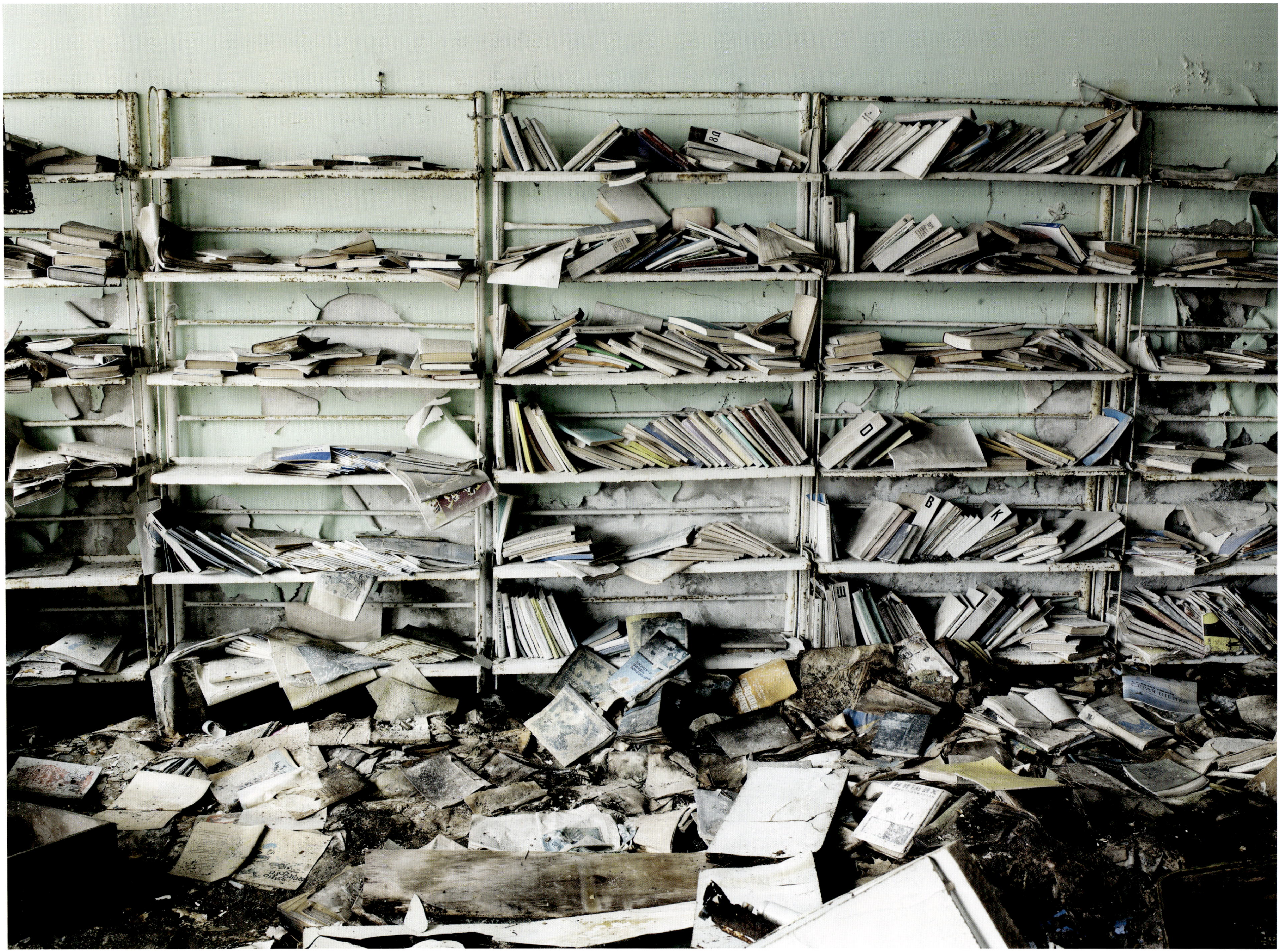

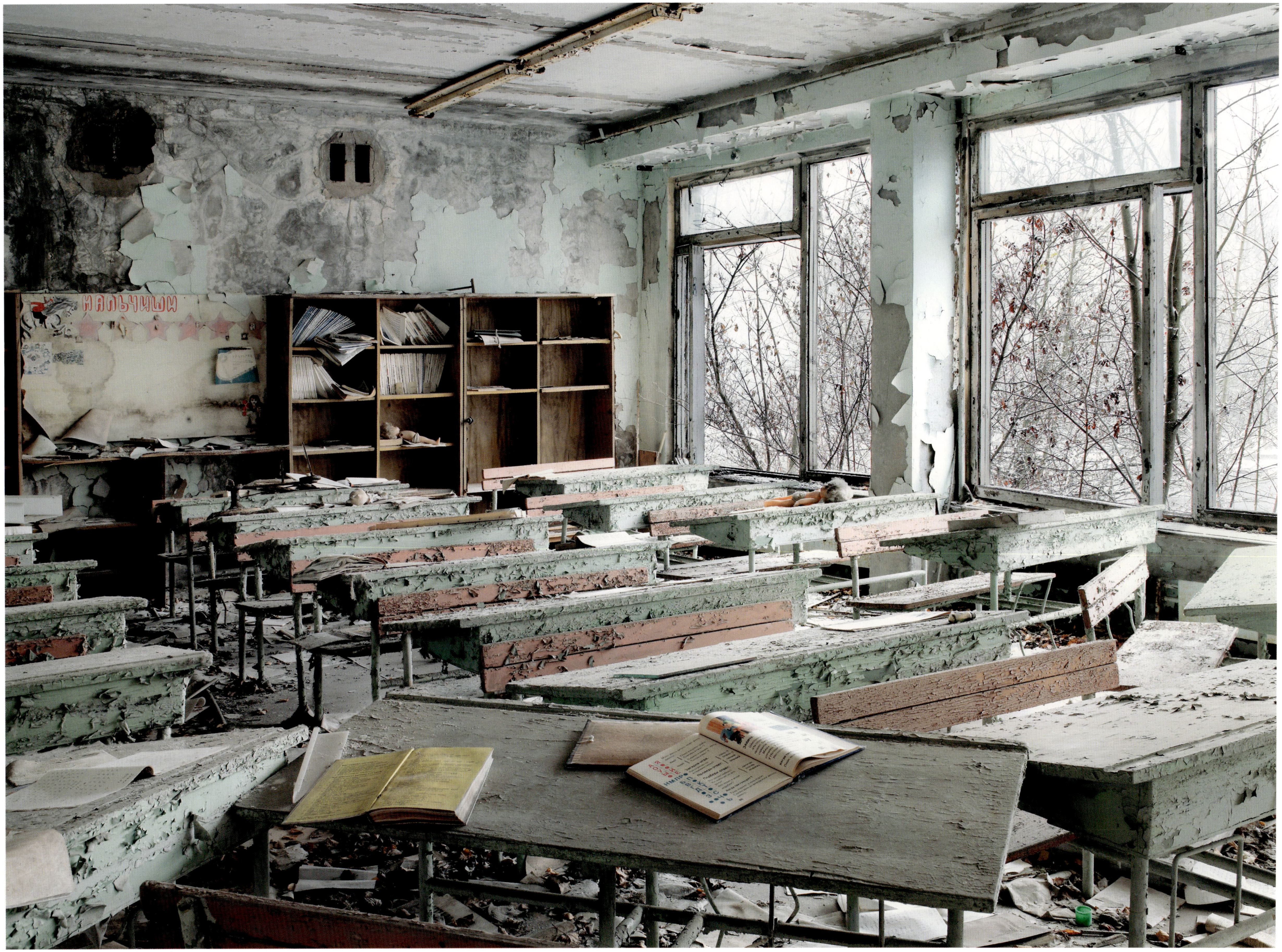

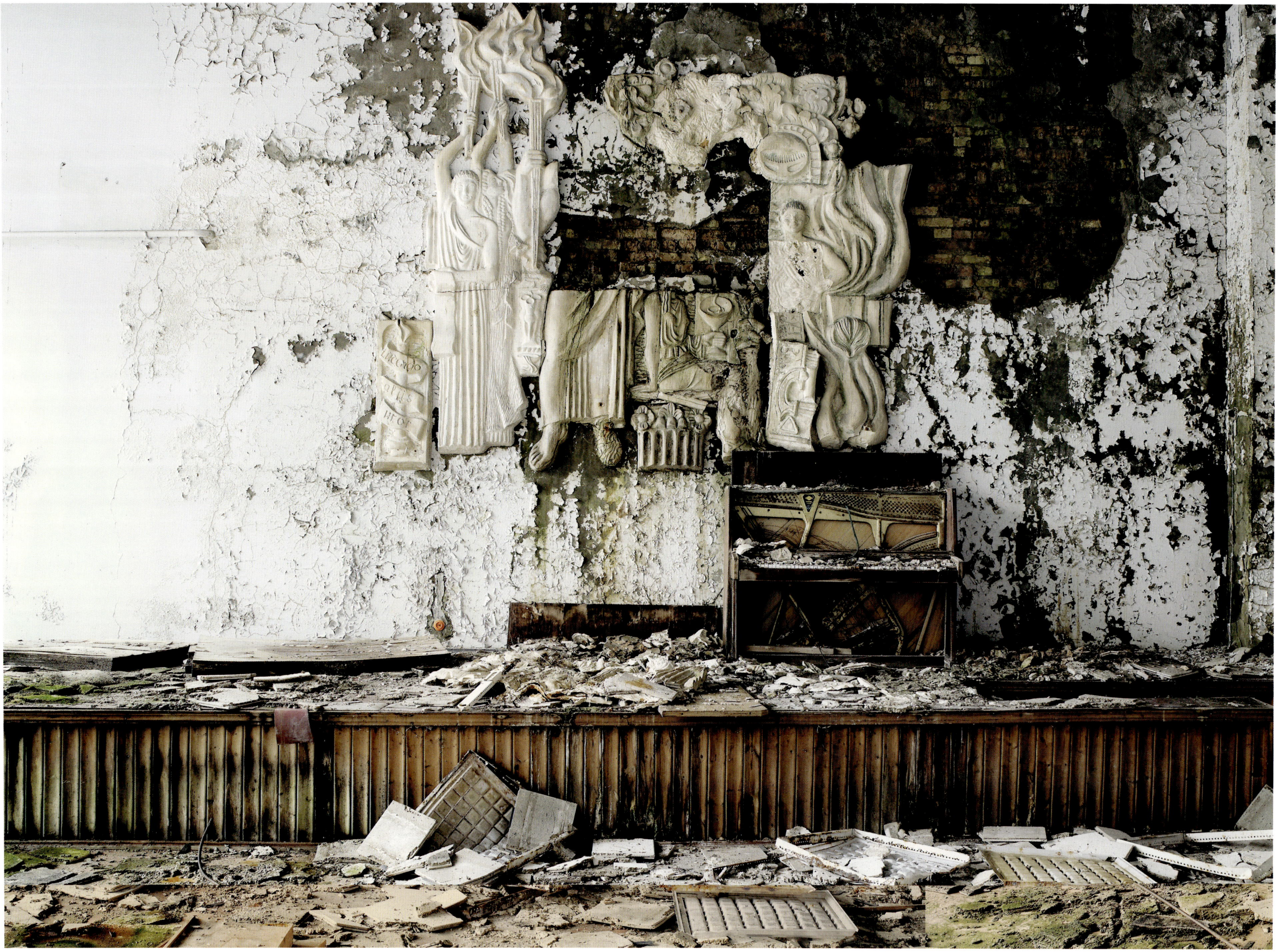

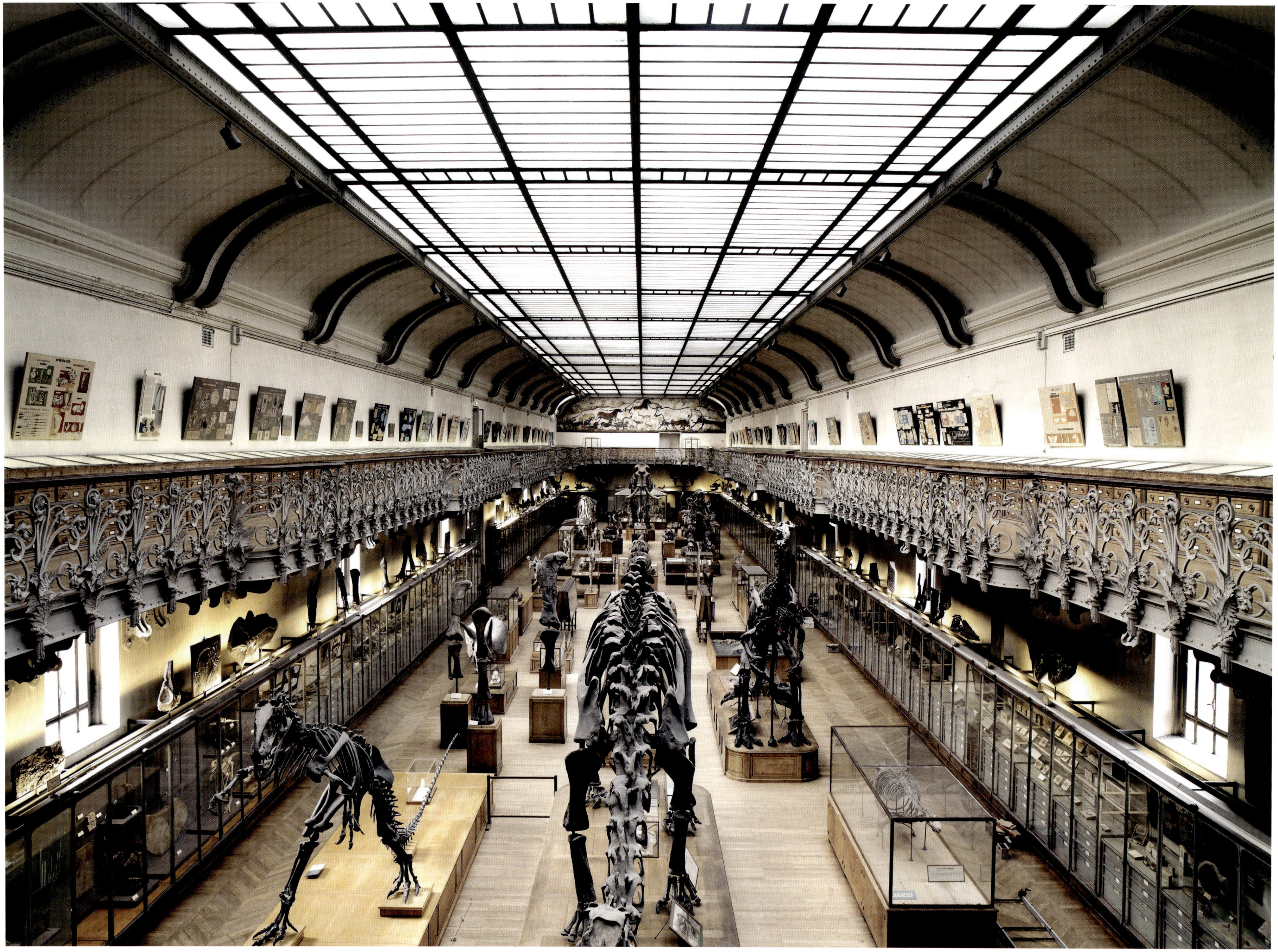

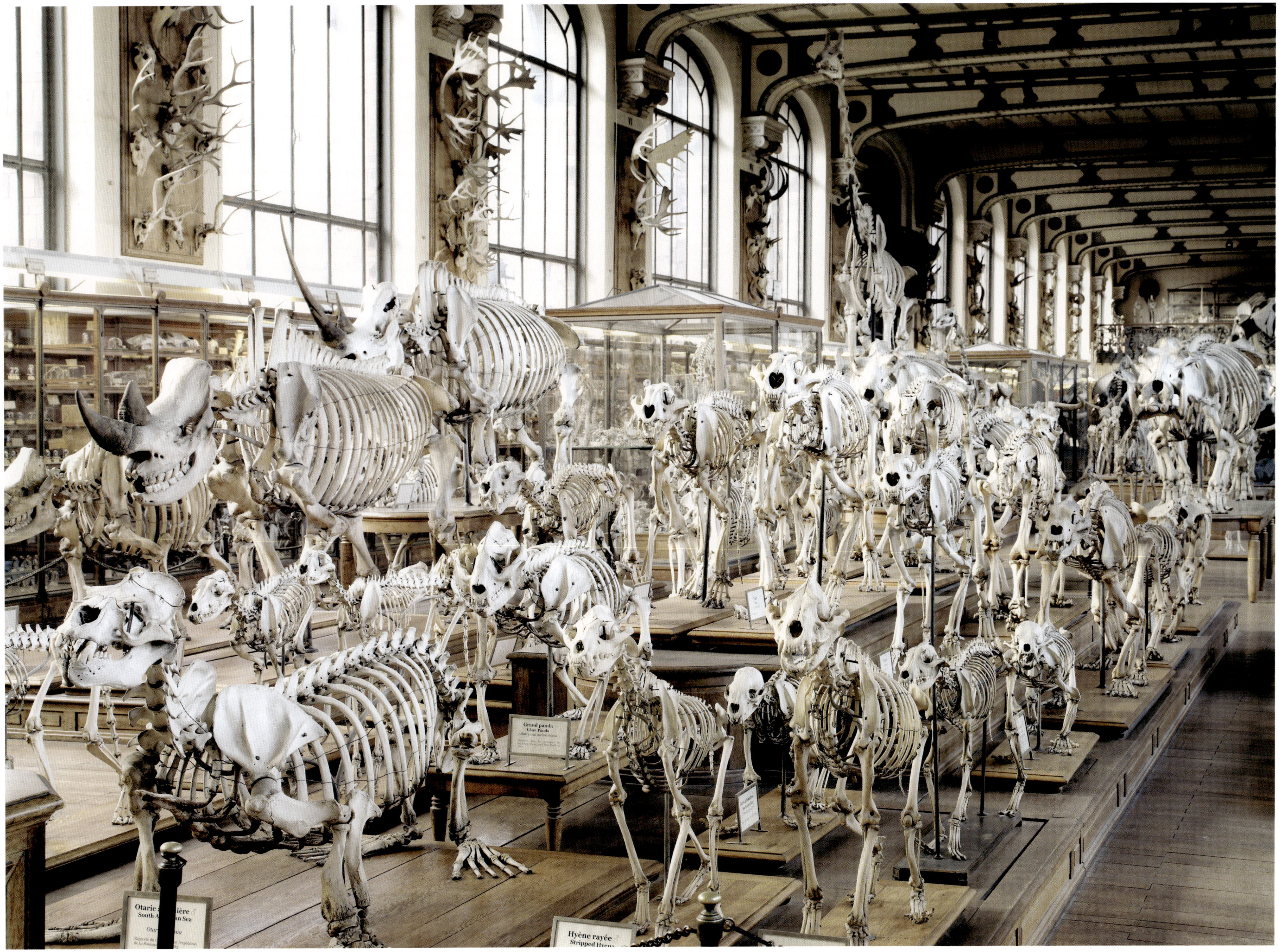
Grand panda
Giant Panda
Hyène rayée

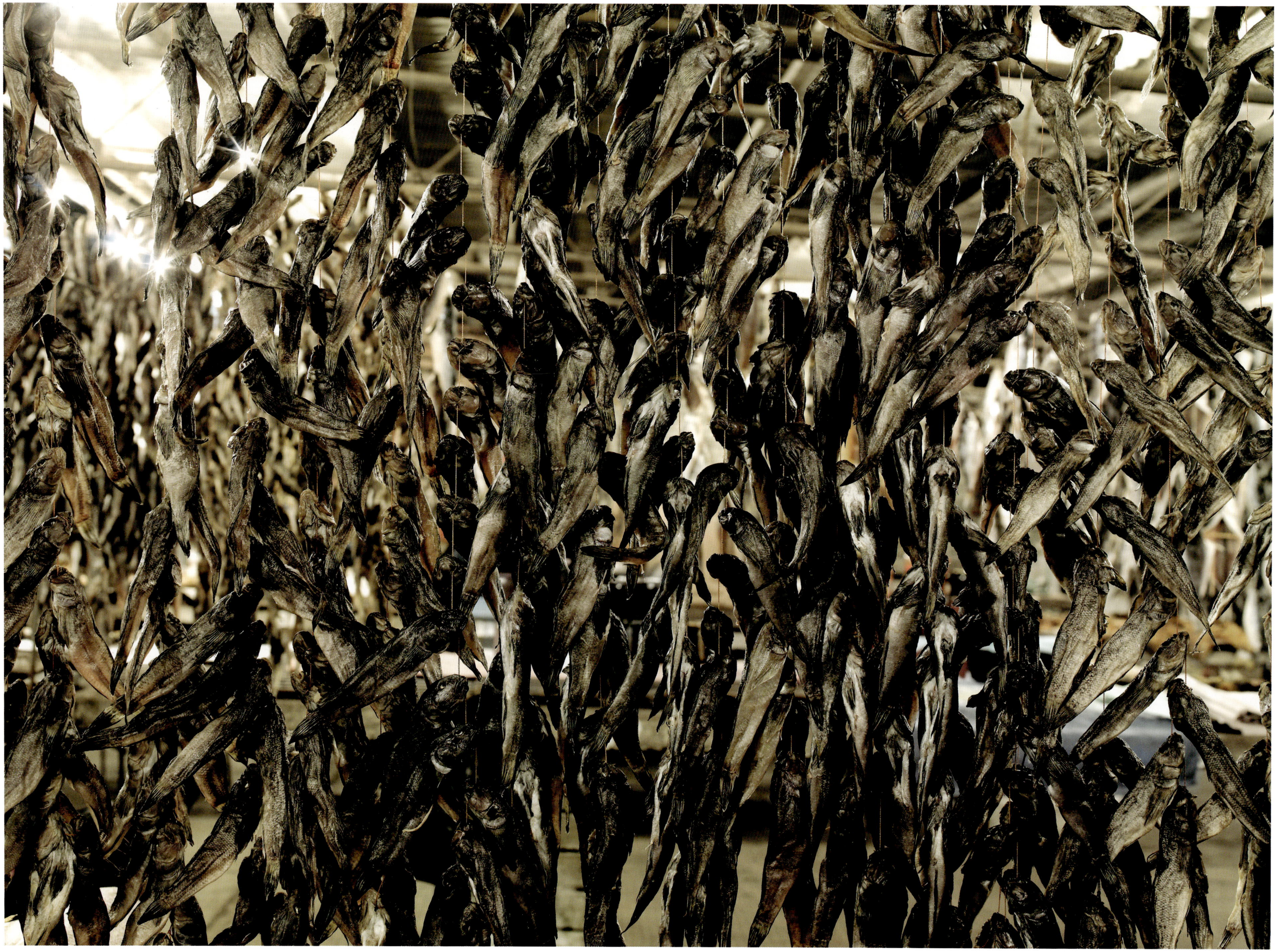

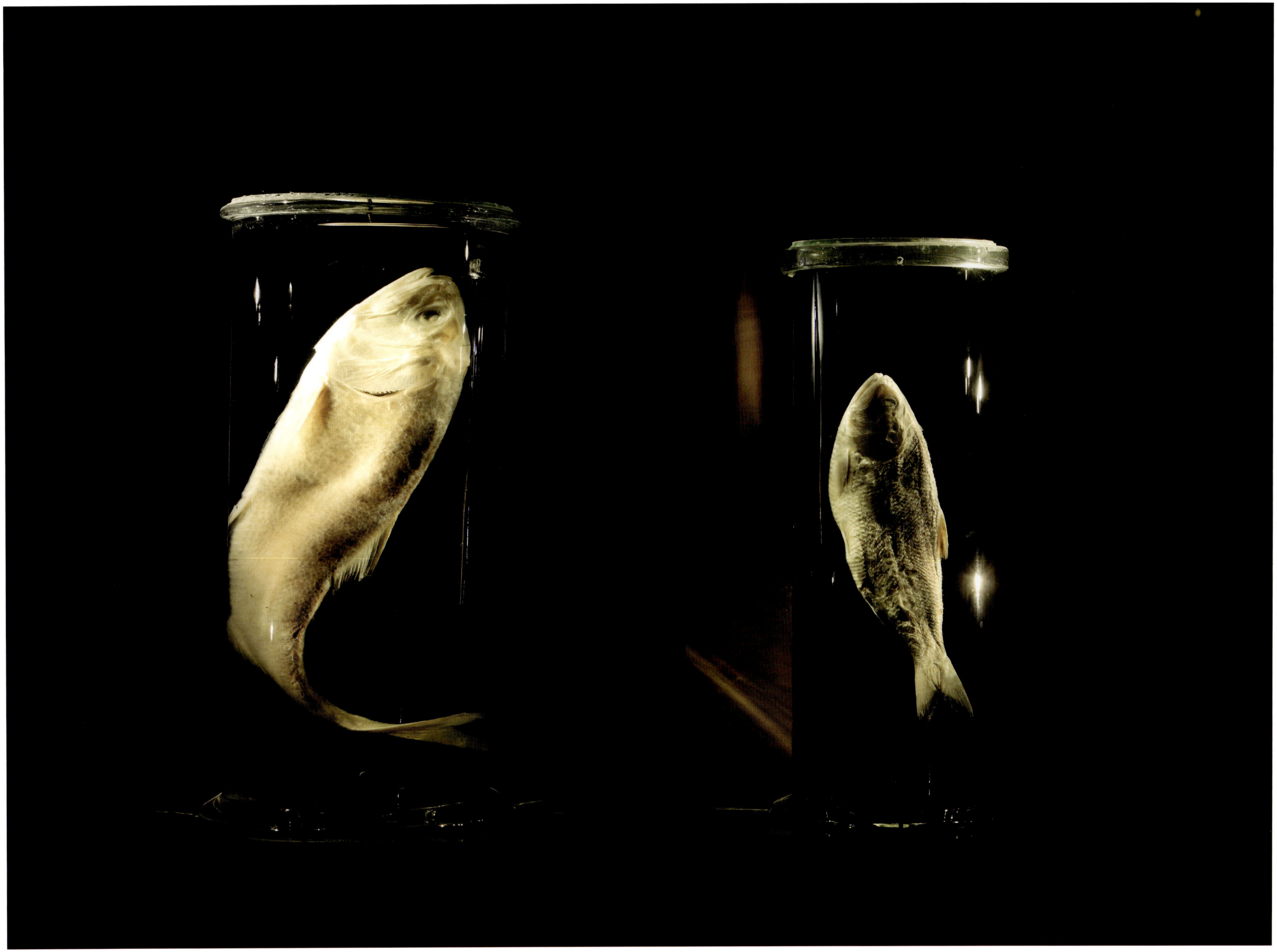

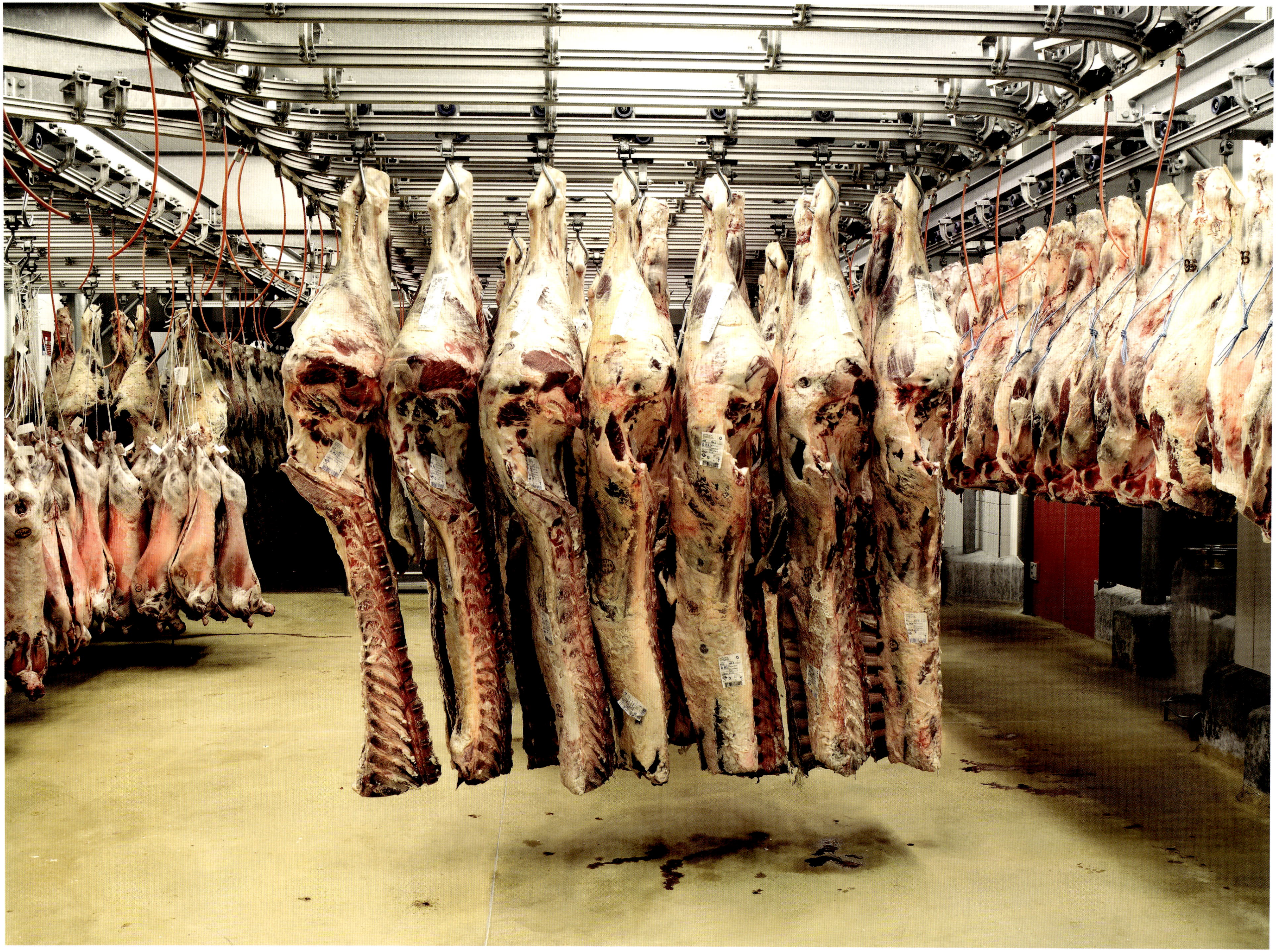

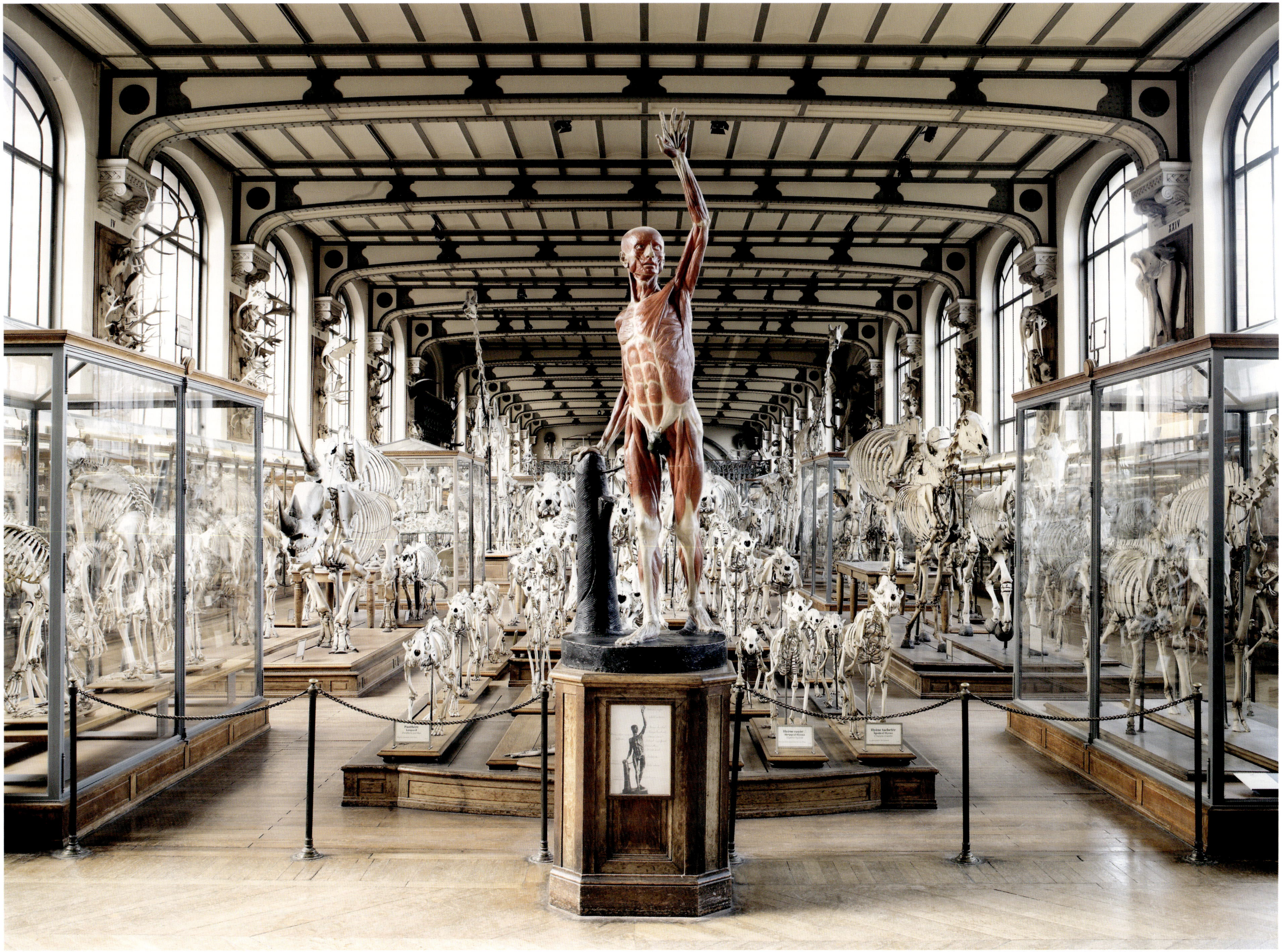

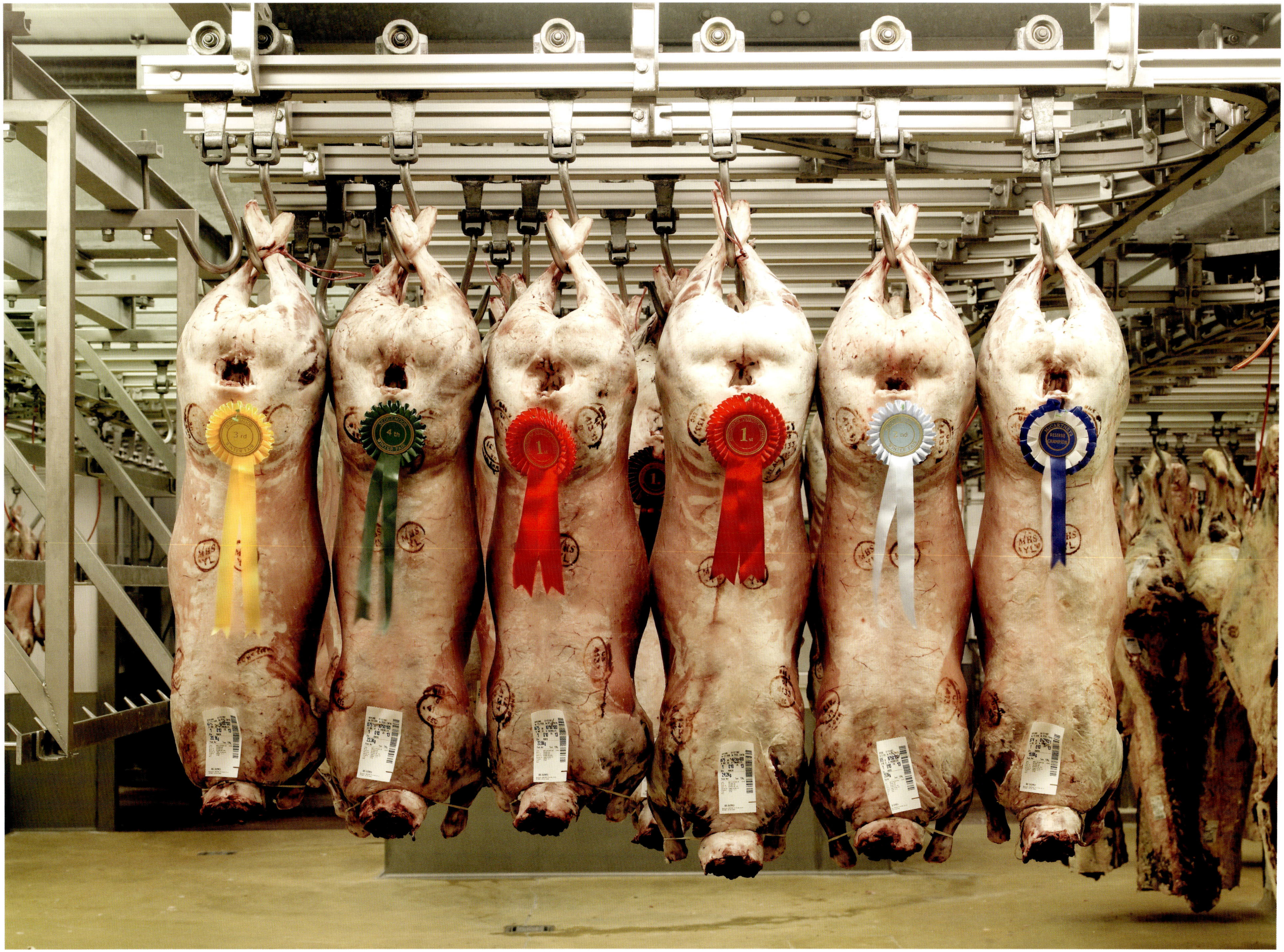

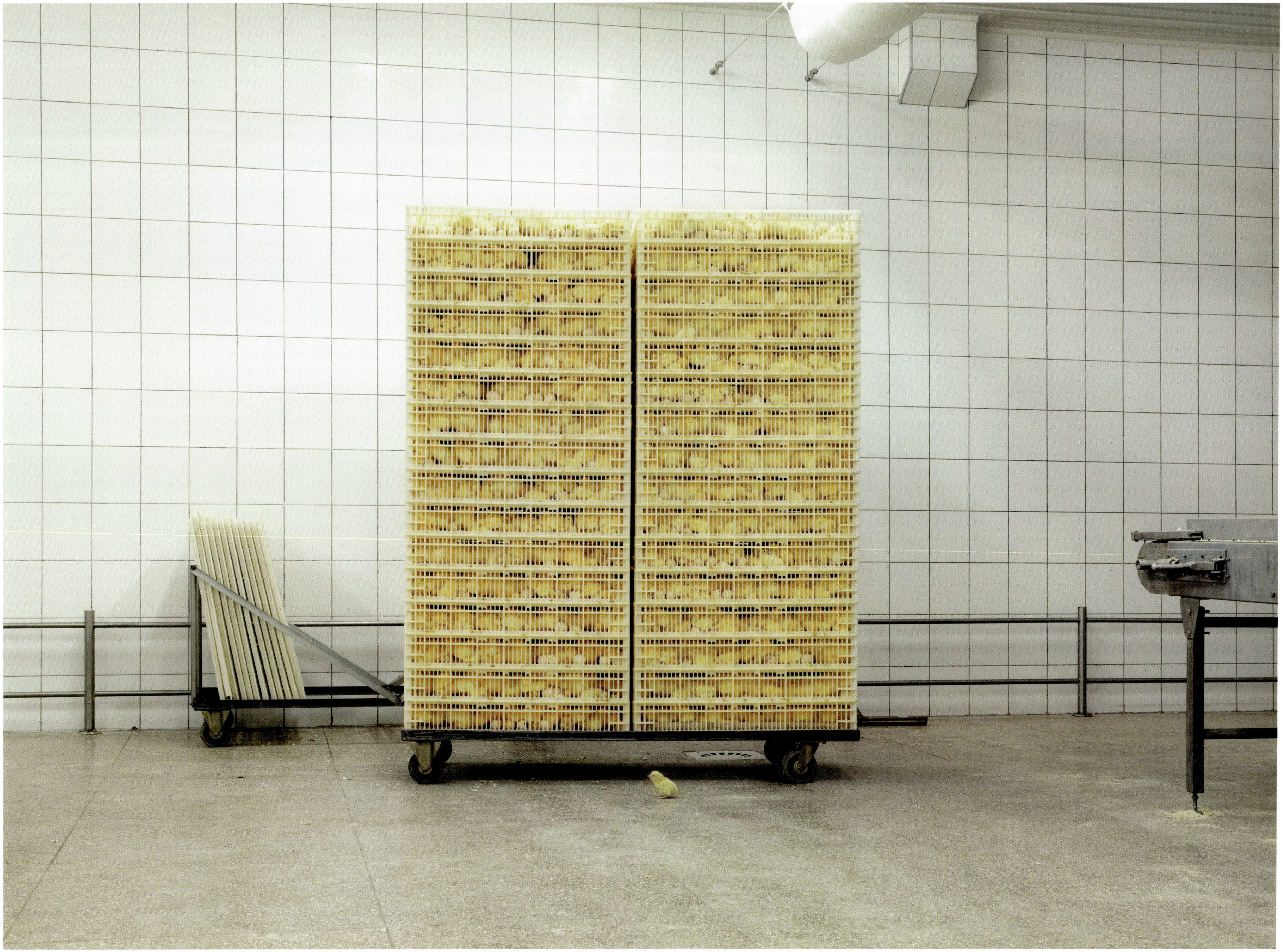

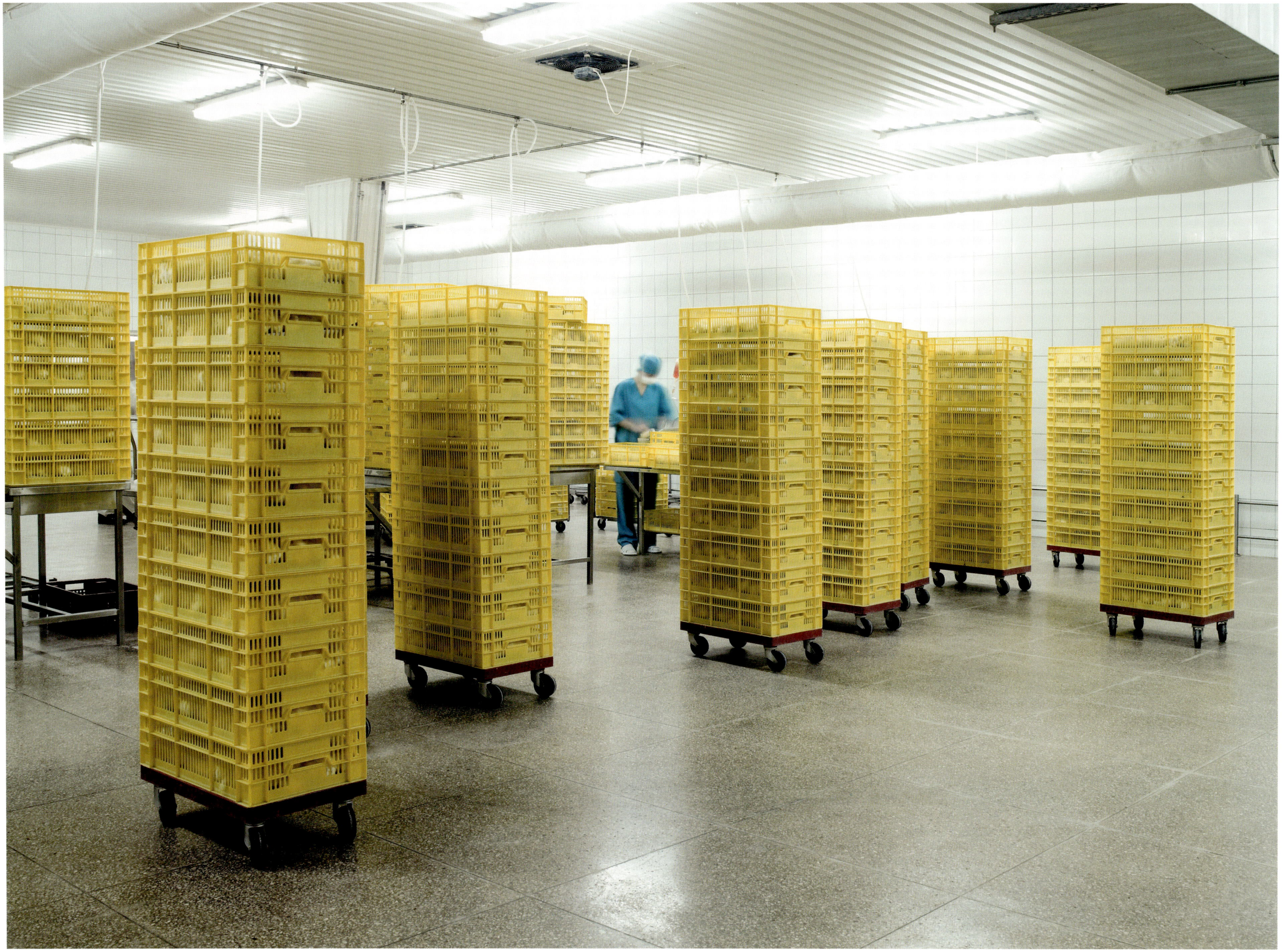

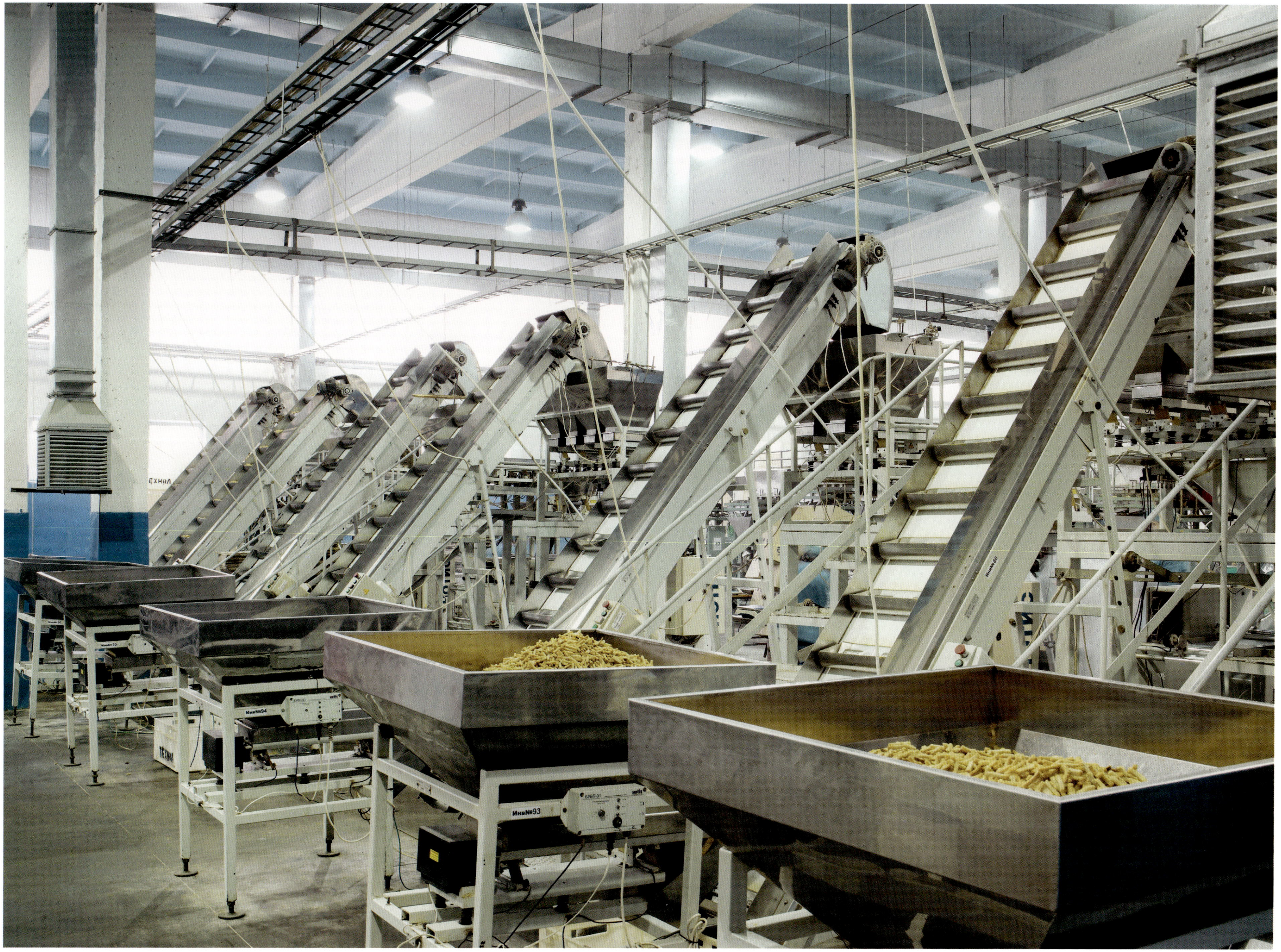

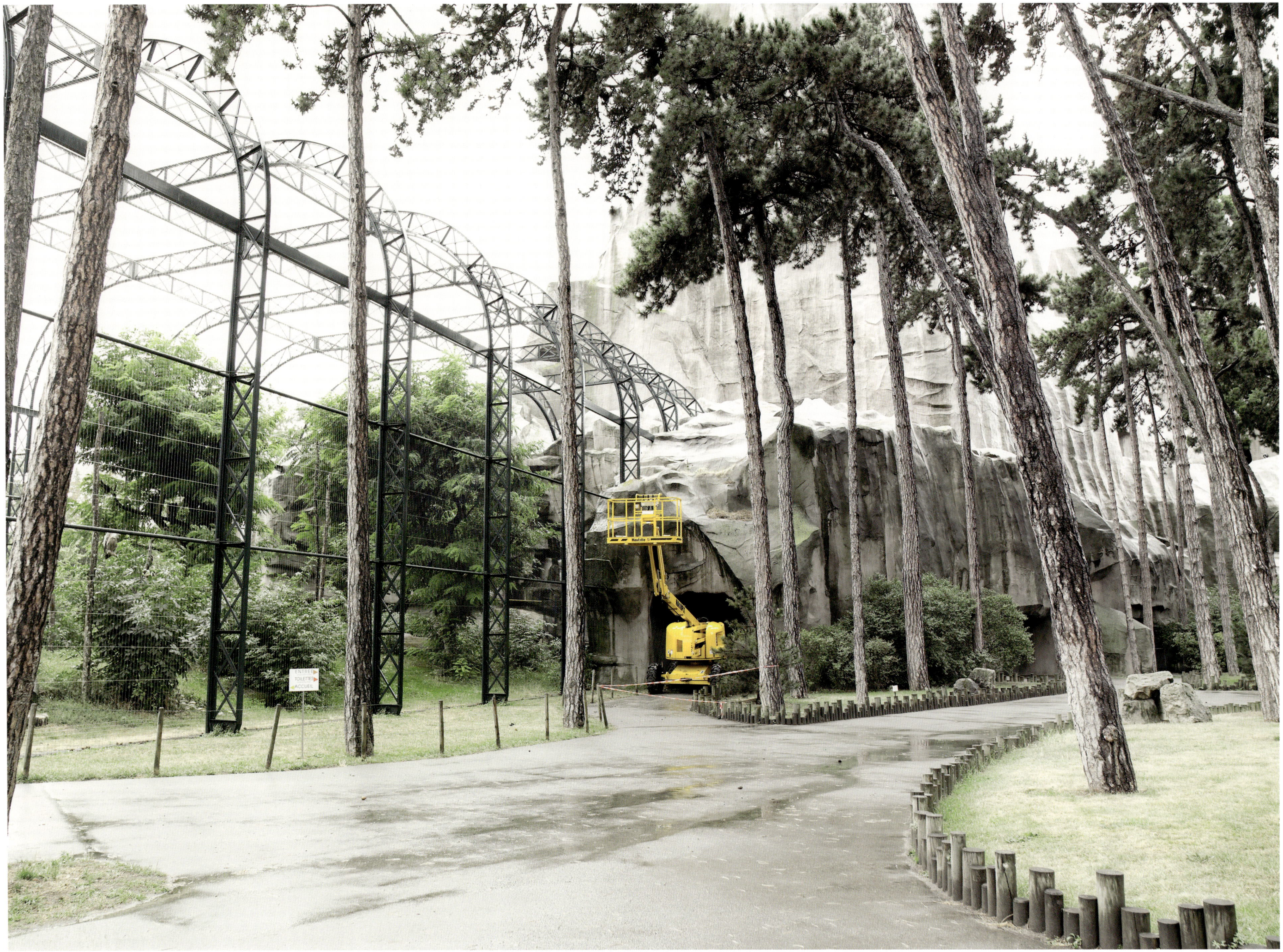
ENTRÉE
TOILETTES
ACCUEIL

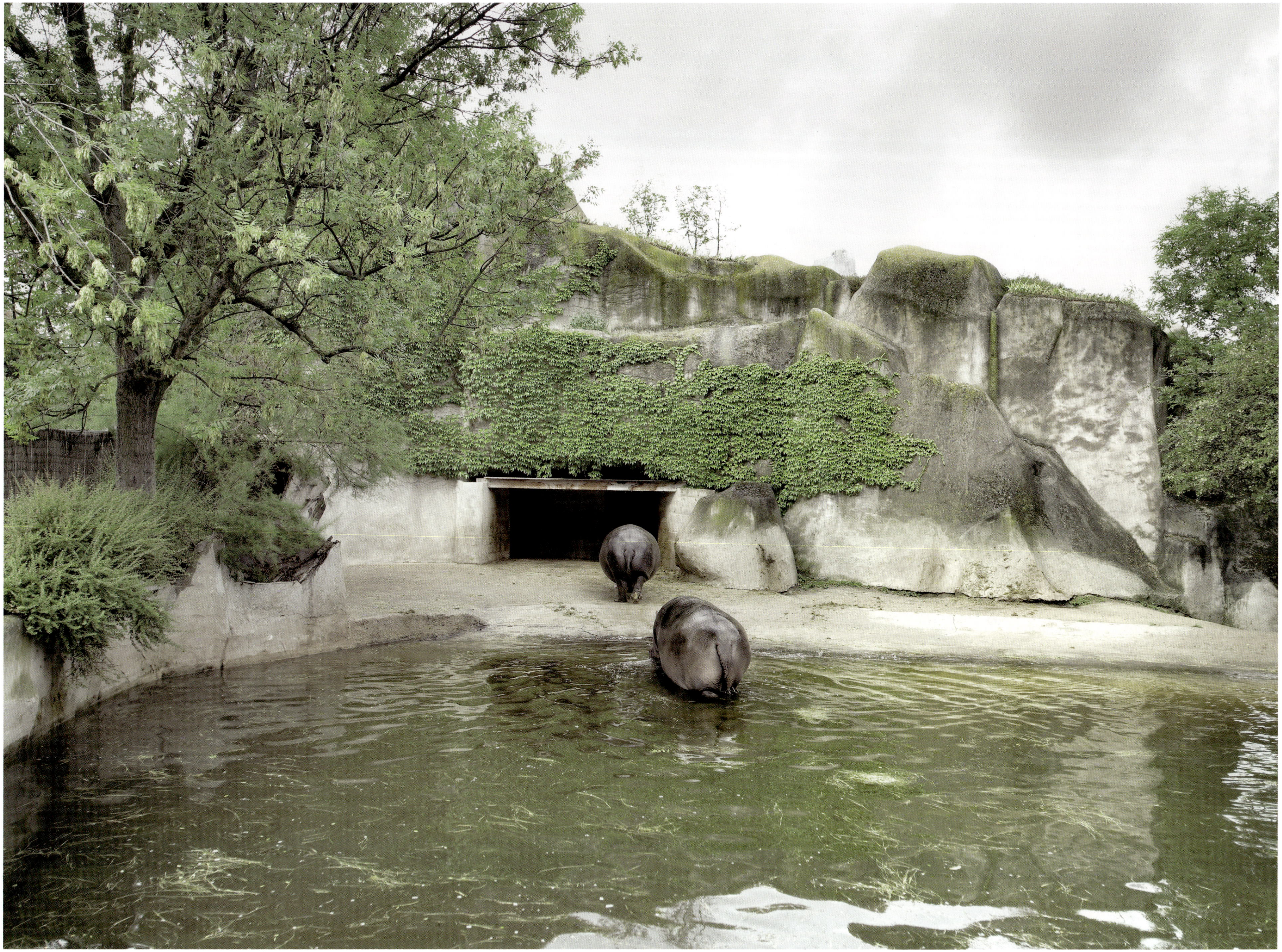

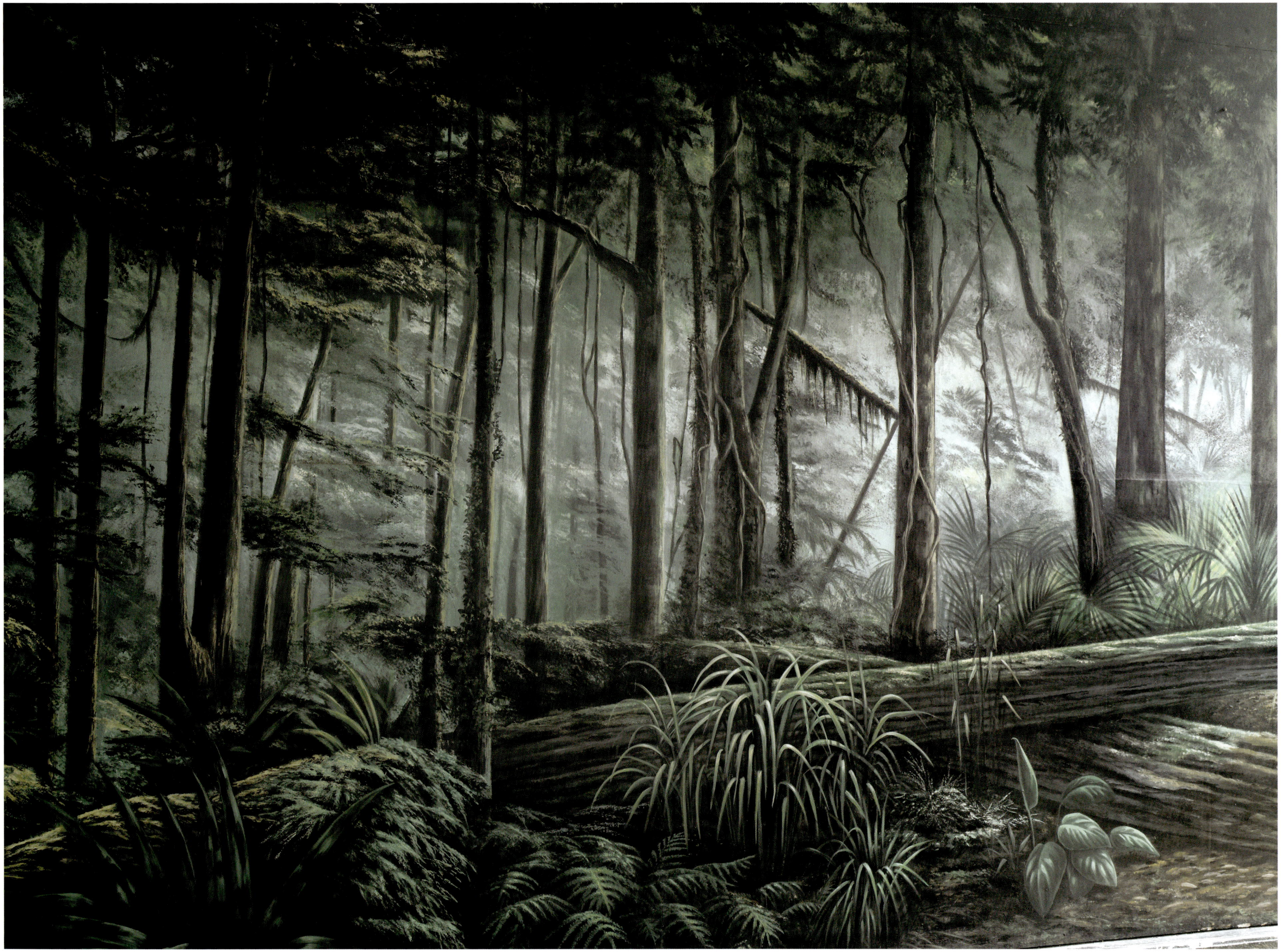

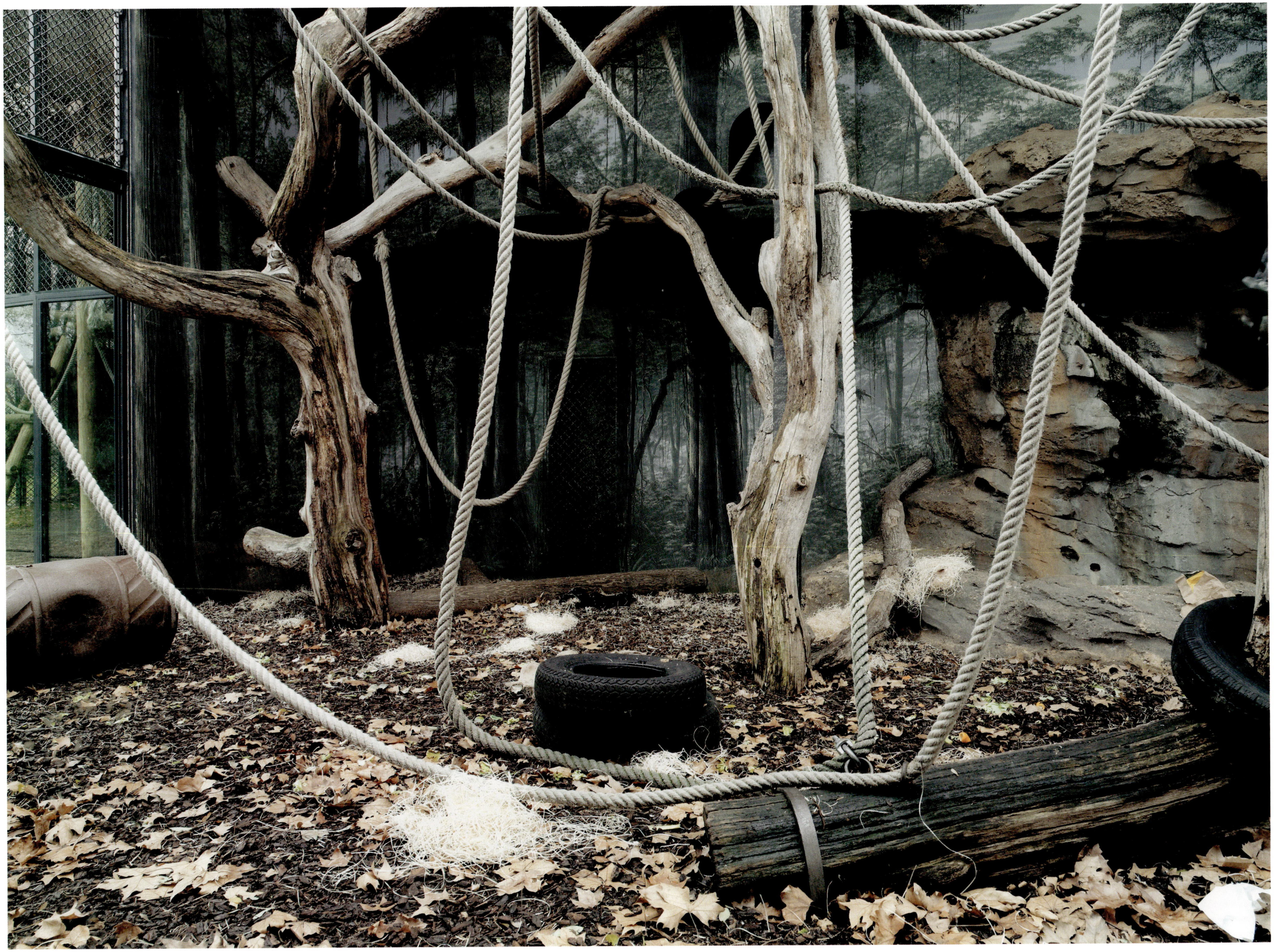

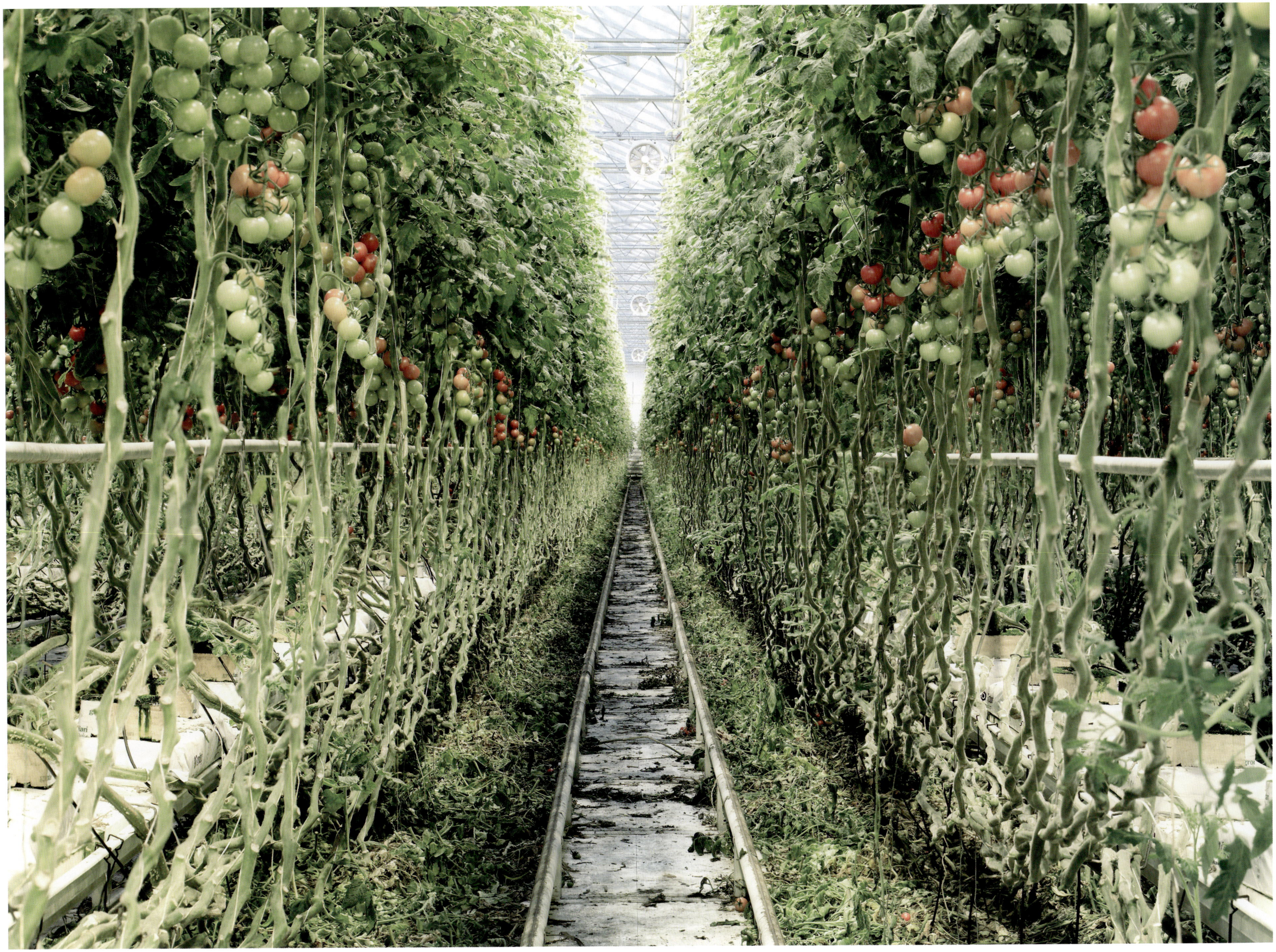

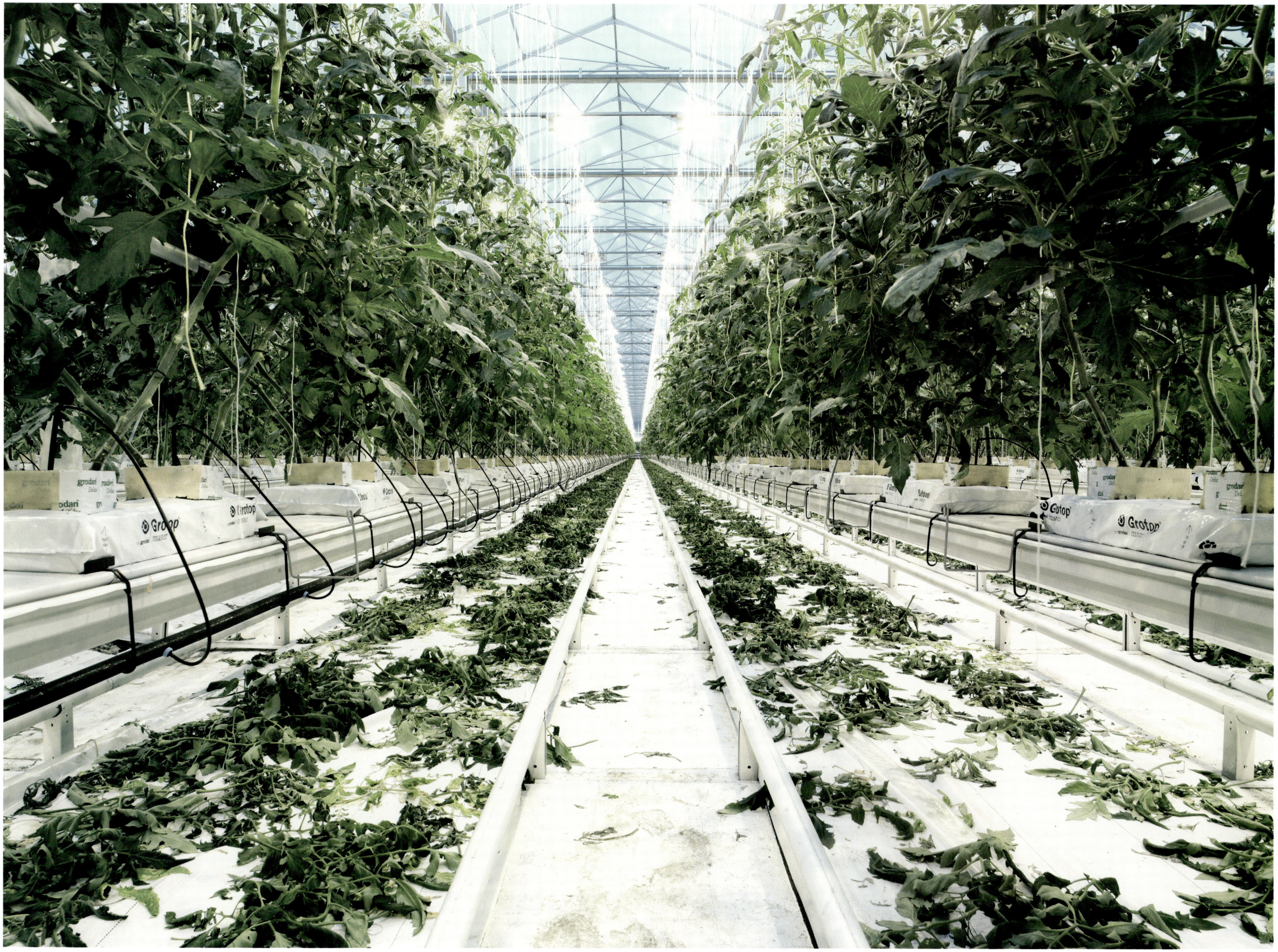
grodan
Grotop

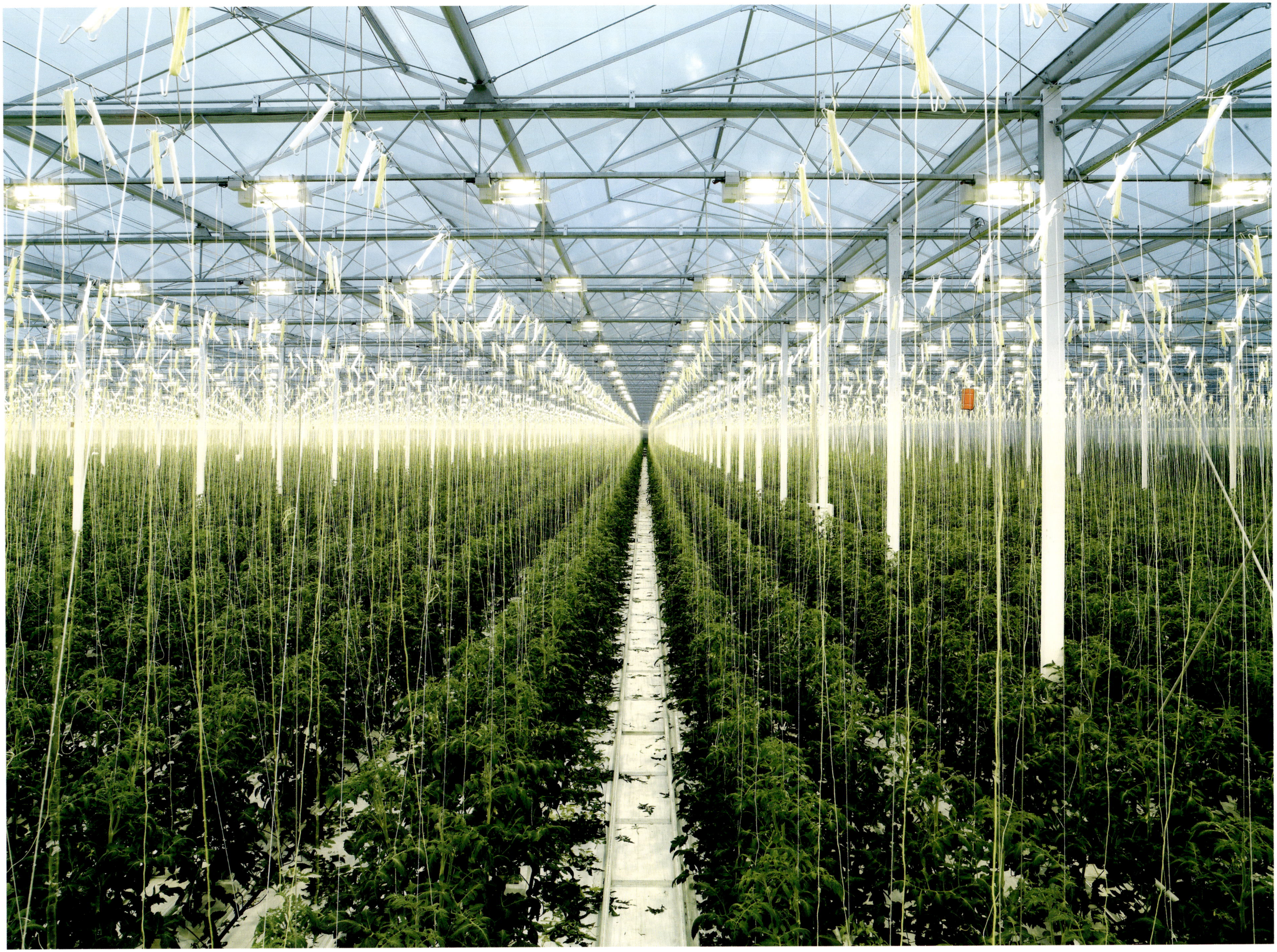

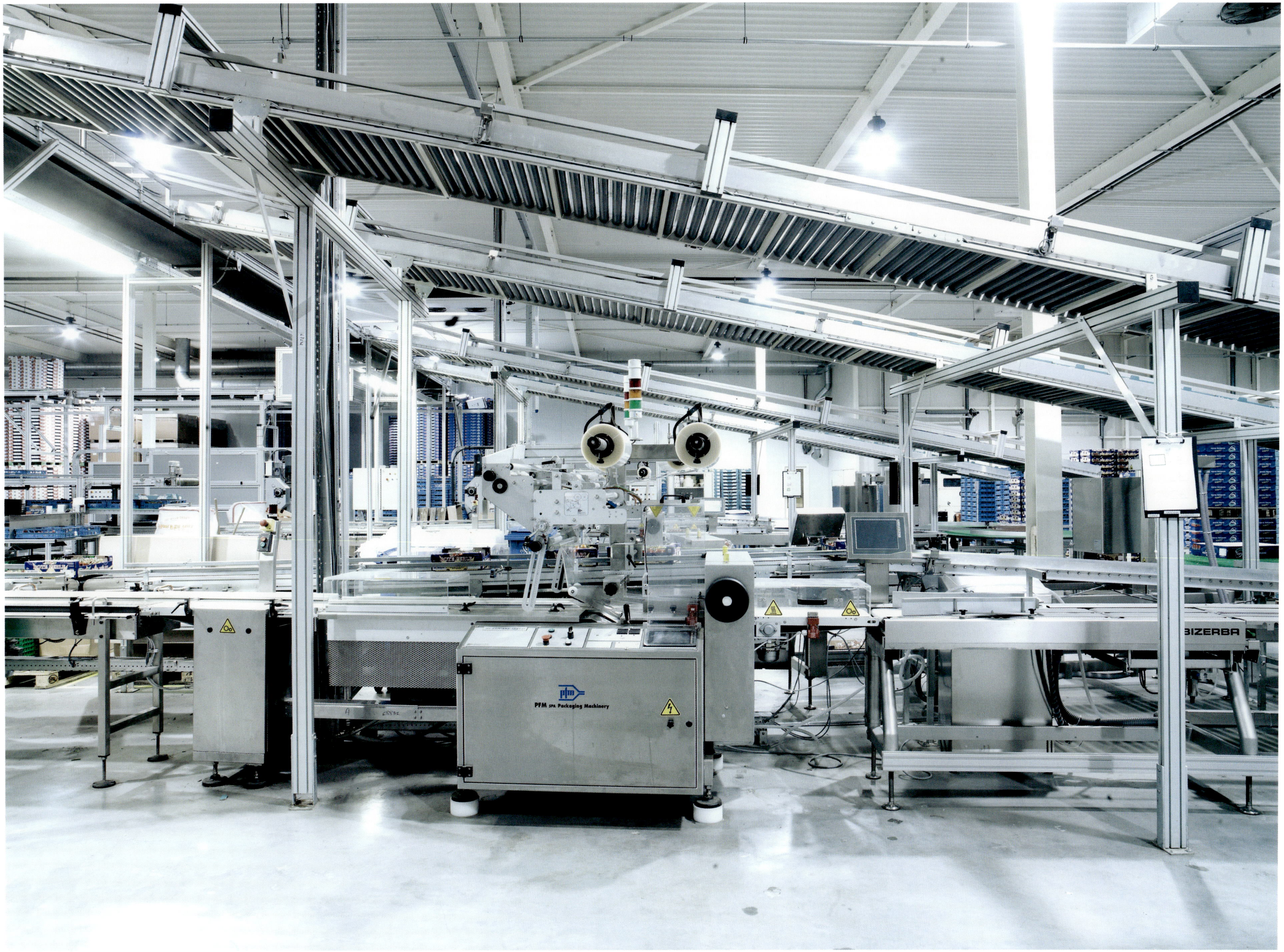
PFM SPA Packaging Machinery
BIZERBA

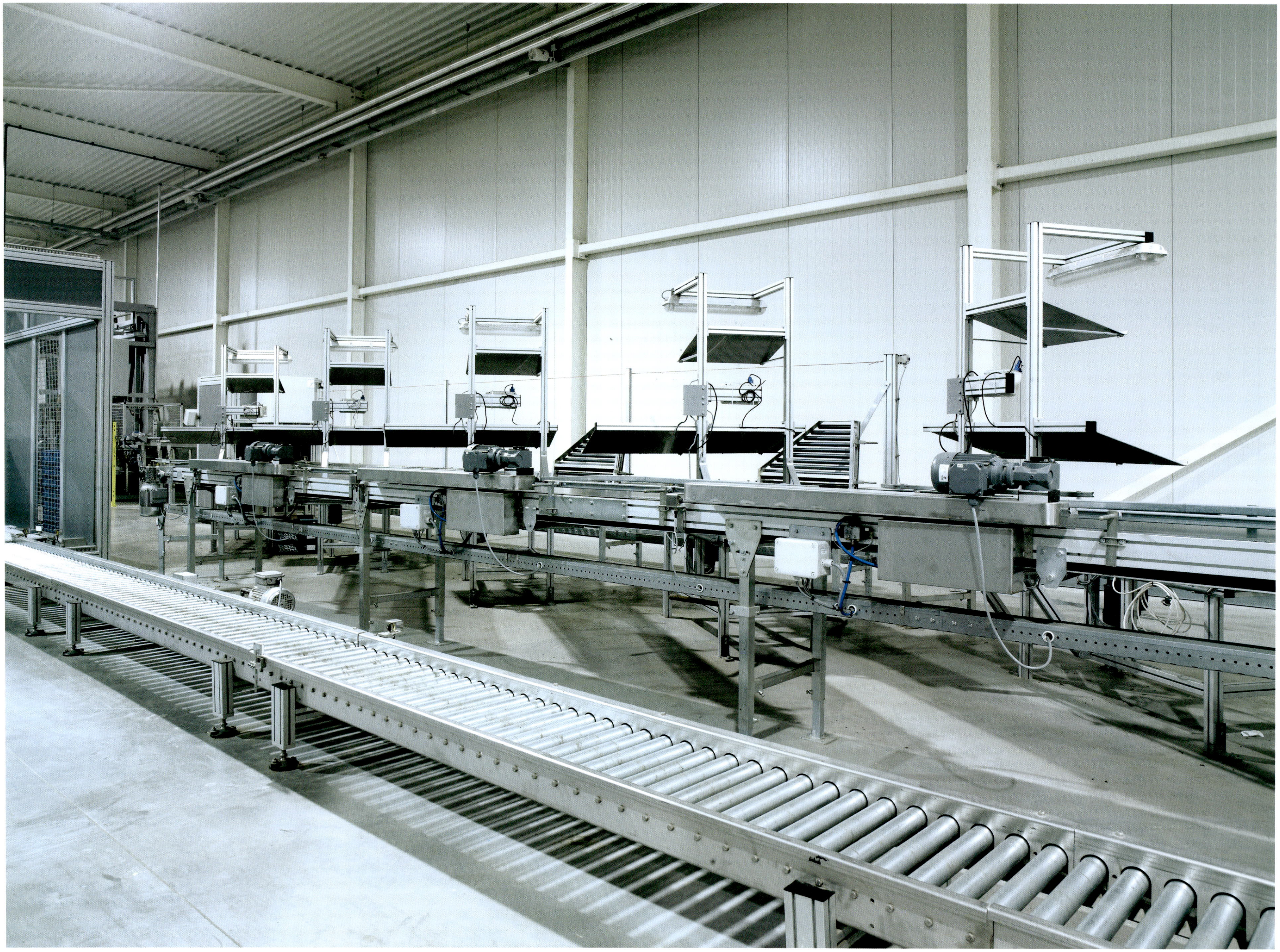

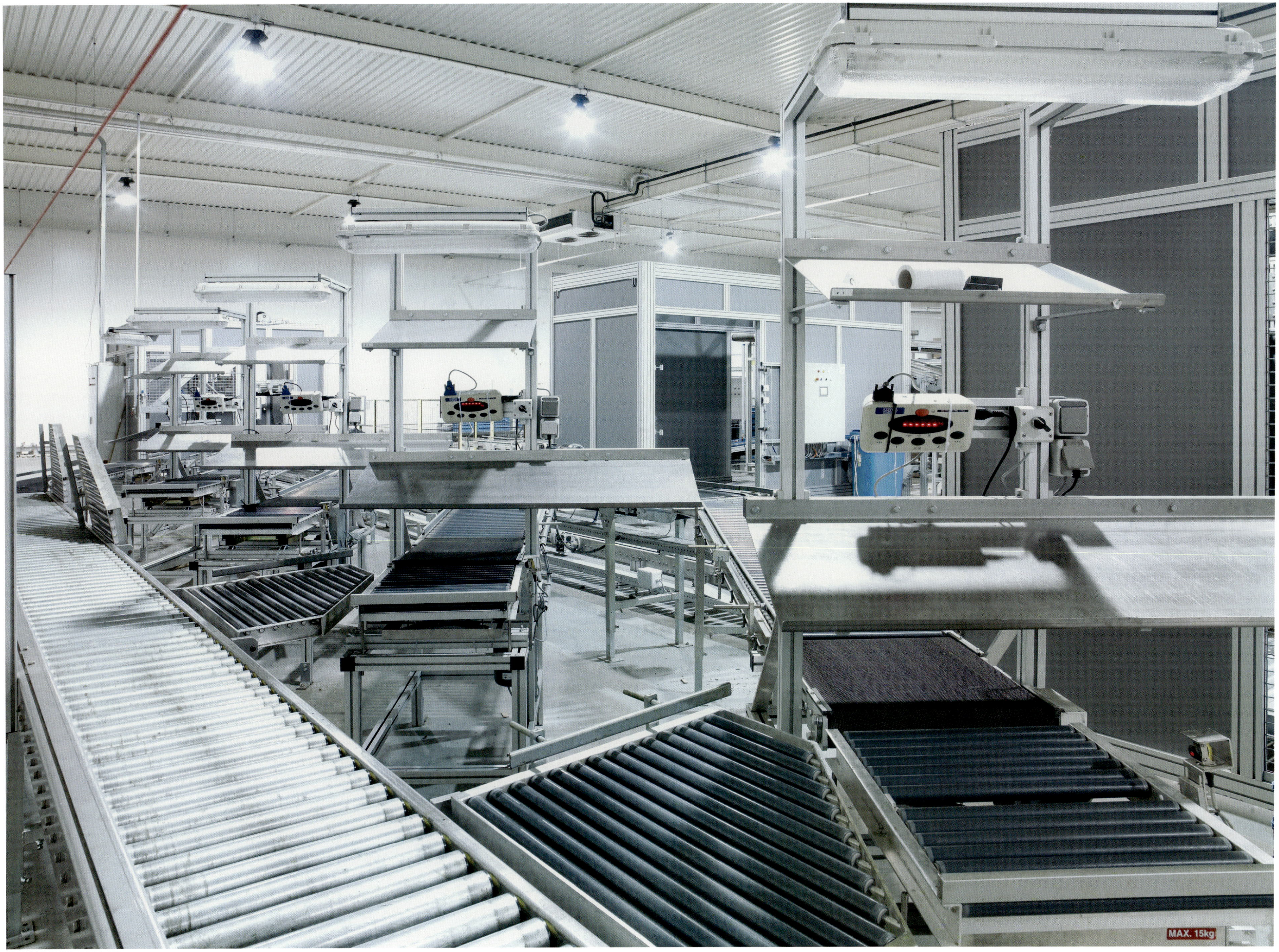
MAX. 15kg

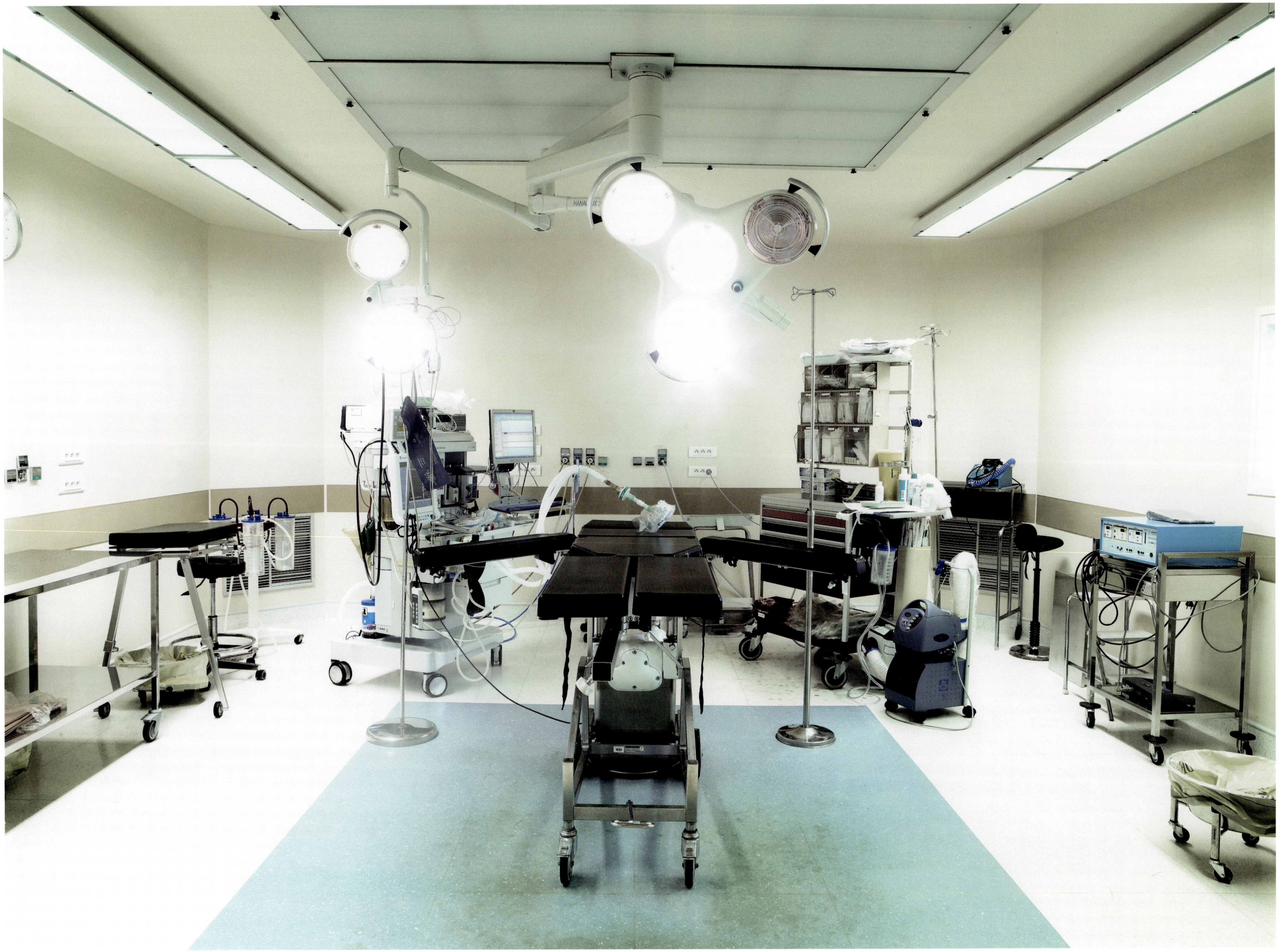

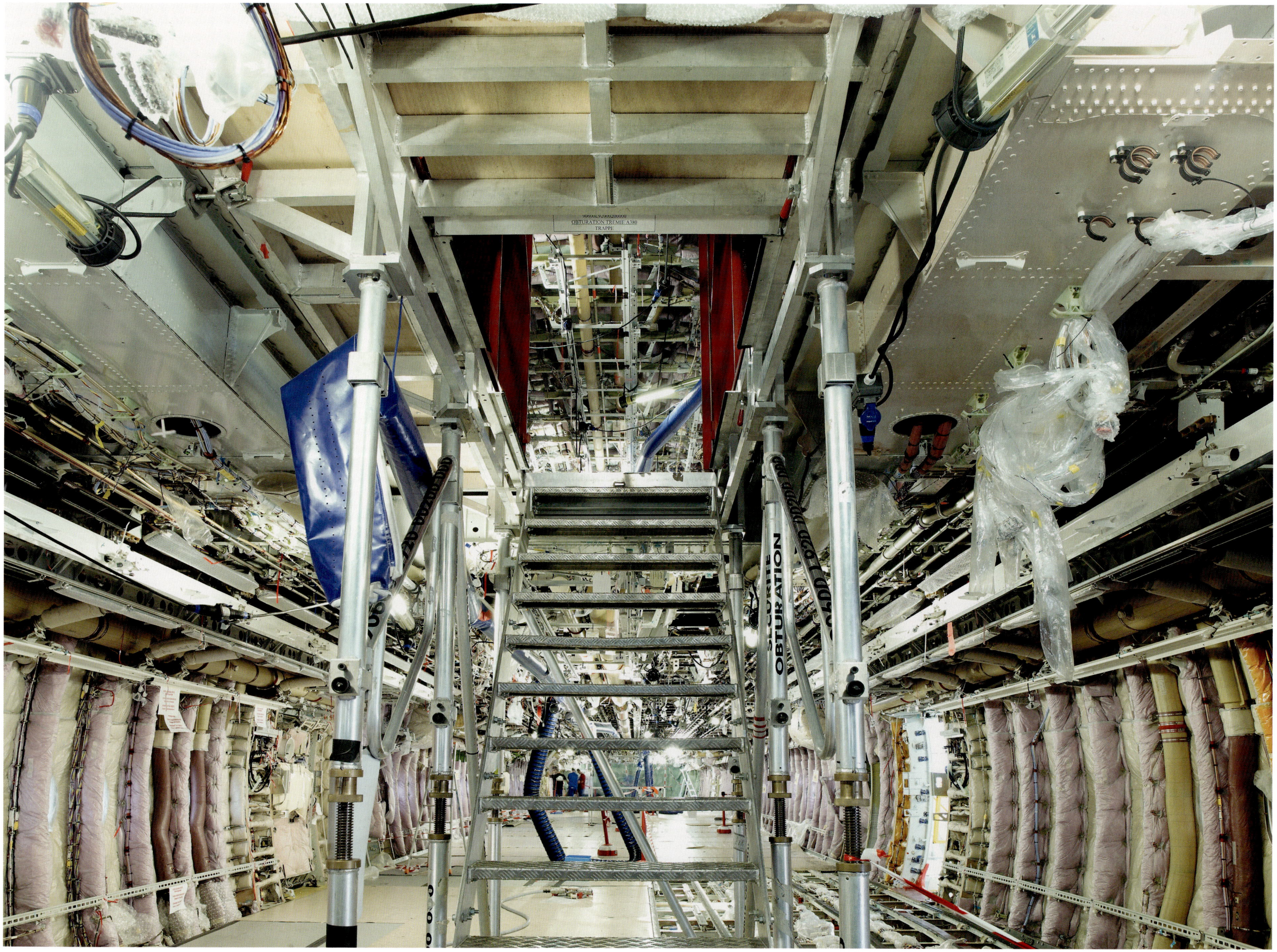
OBTURATION TREMIE A380
TRAPPE
SECURITE
OBTURATION

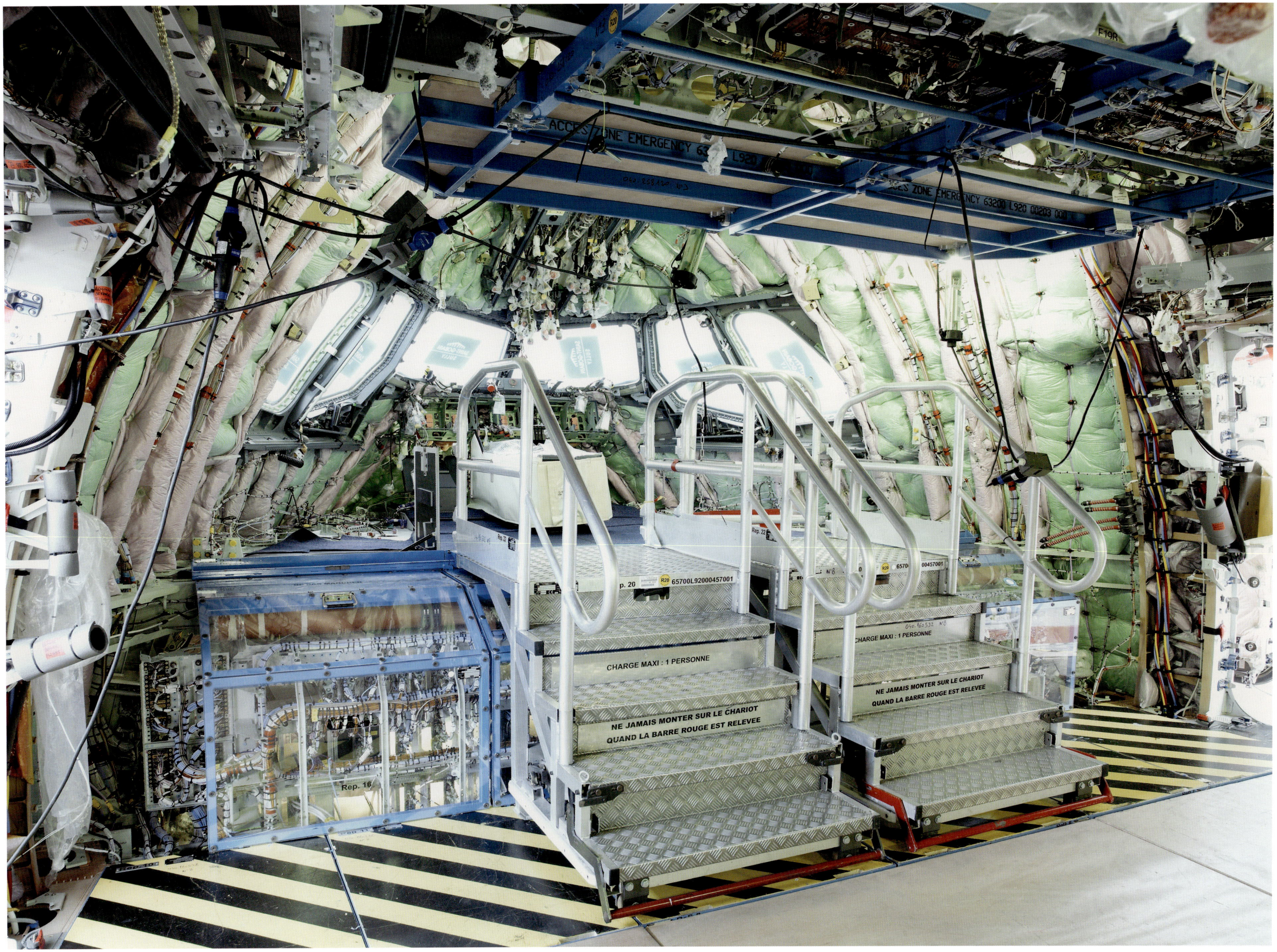
CHARGE MAXI : 1 PERSONNE
NE JAMAIS MONTER SUR LE CHARIOT
QUAND LA BARRE ROUGE EST RELEVEE
CHARGE MAXI : 1 PERSONNE
NE JAMAIS MONTER SUR LE CHARIOT
QUAND LA BARRE ROUGE EST RELEVEE
Rep. 16
65700L92000457001

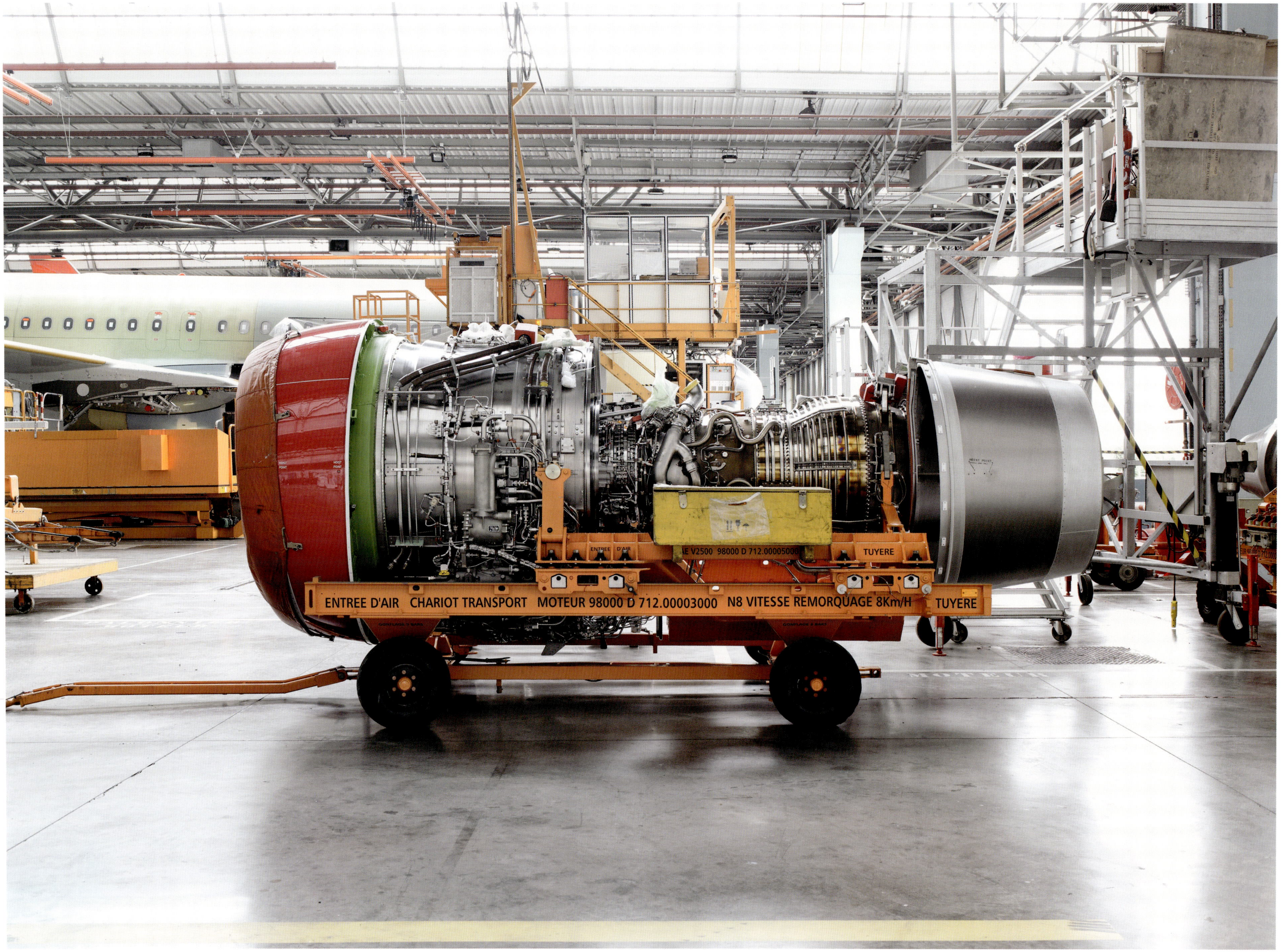
ENTREE D'AIR CHARIOT TRANSPORT MOTEUR 98000 D 712.00003000 N8 VITESSE REMORQUAGE 8Km/H TUYERE
TUYERE

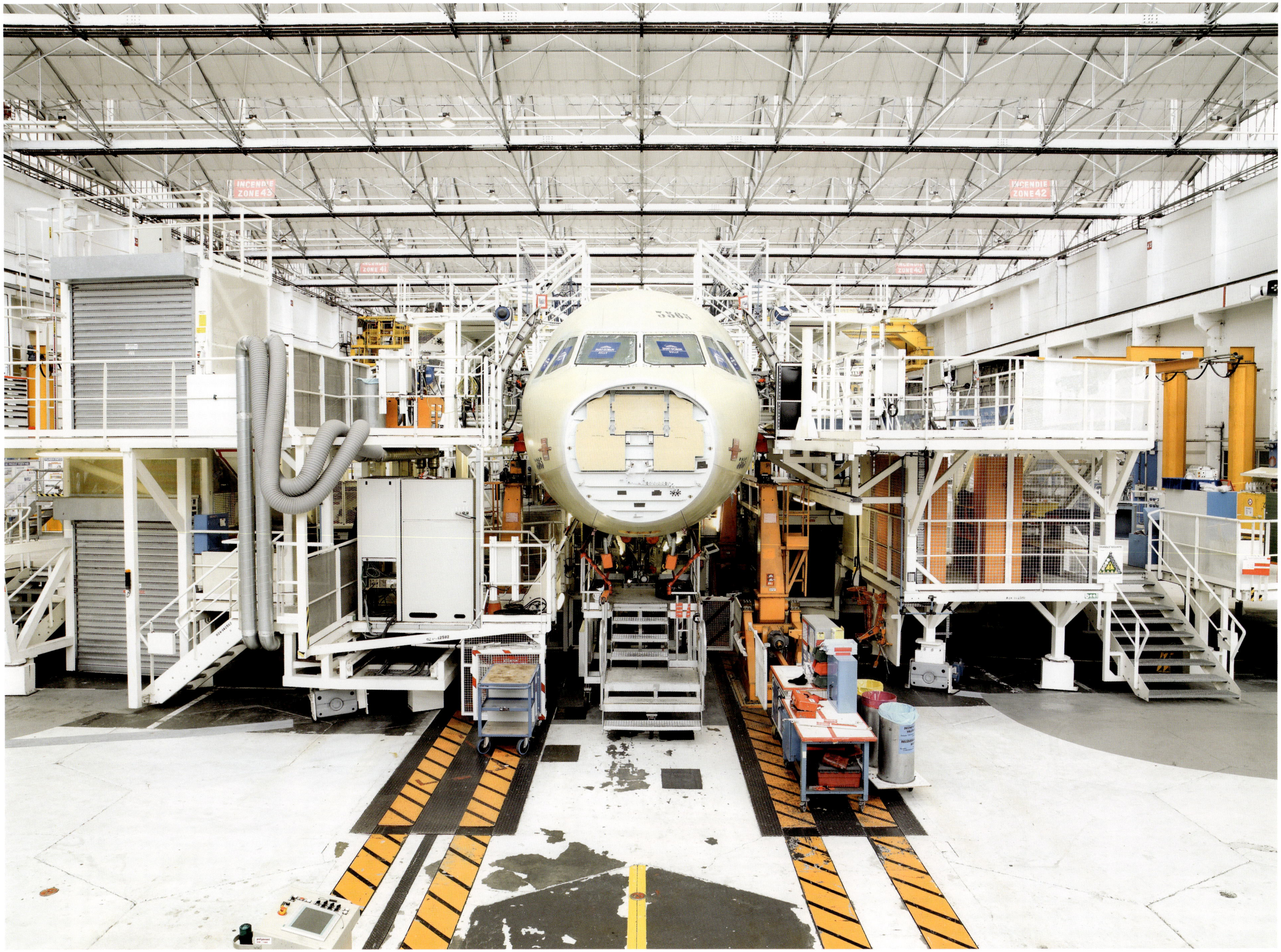
INCENDIE ZONE 43
INCENDIE ZONE 42

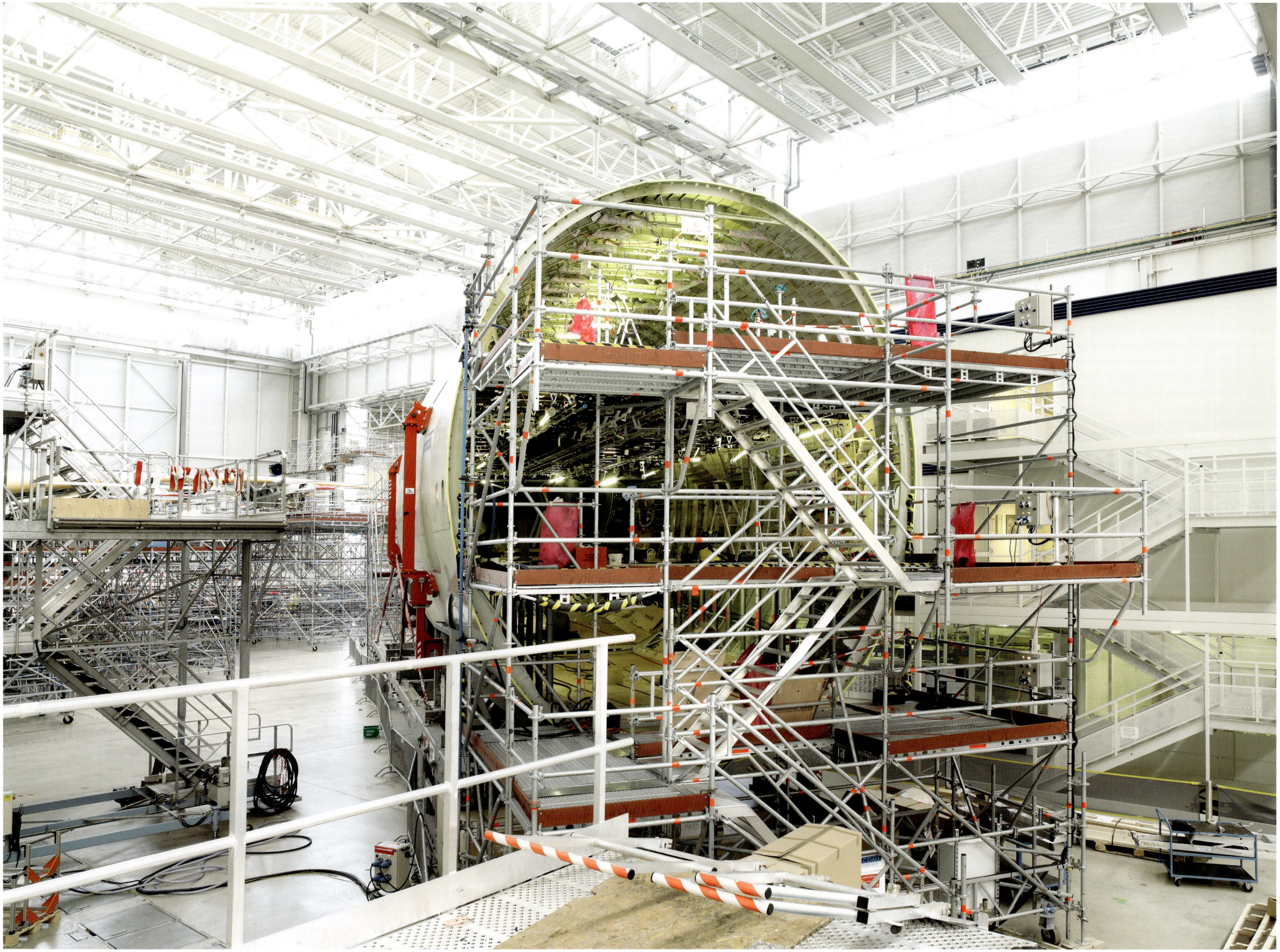

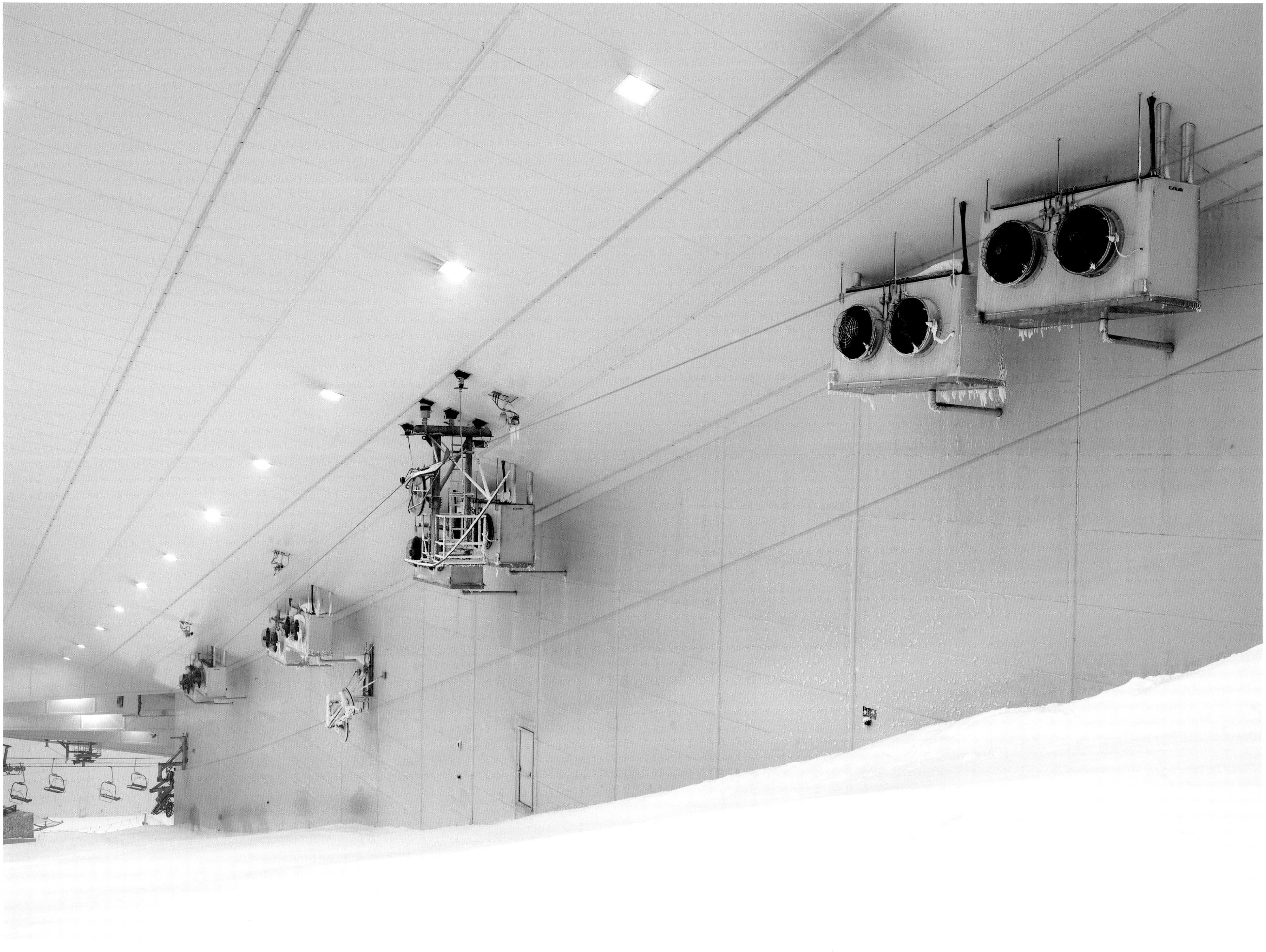

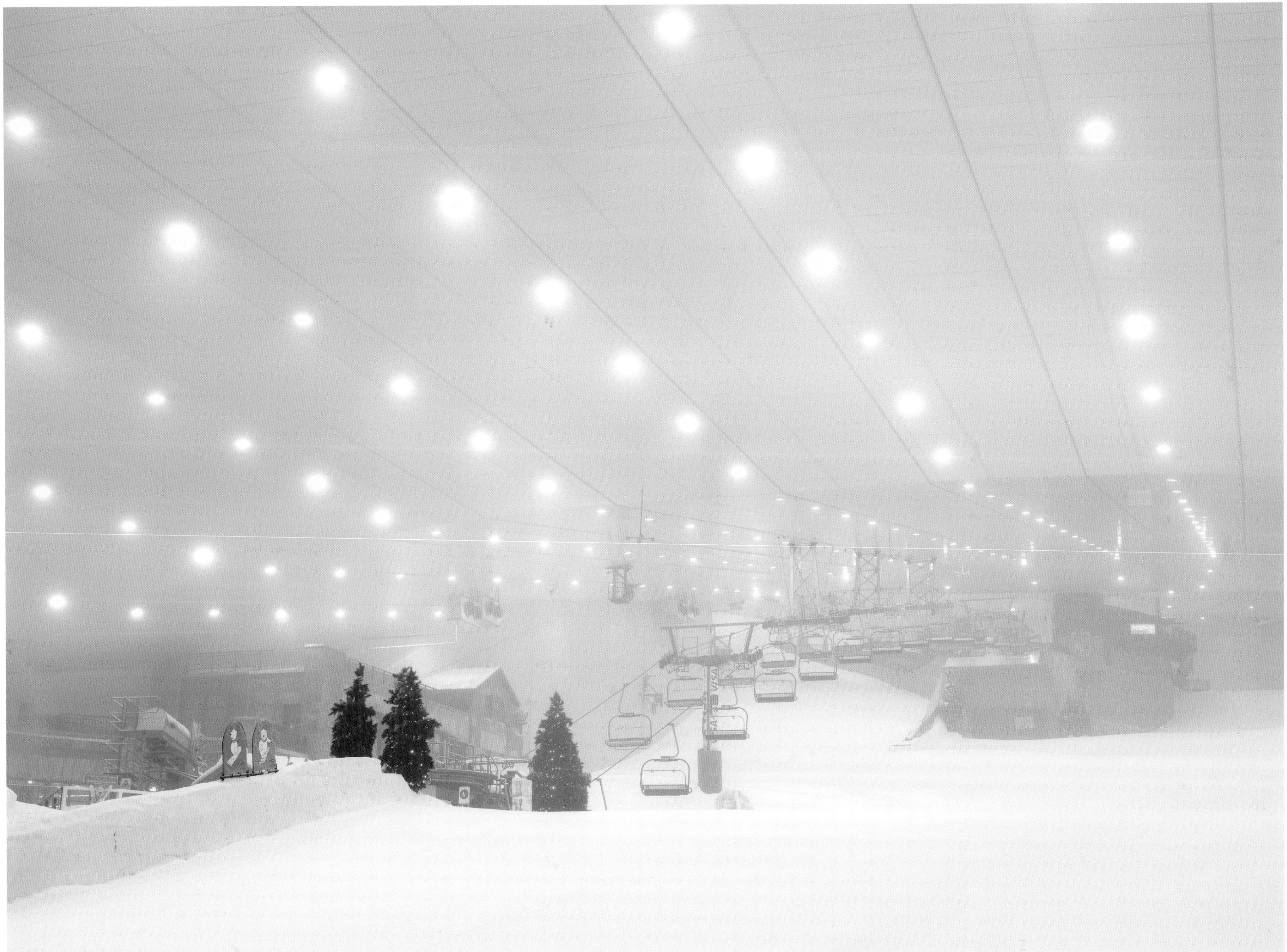
VISA

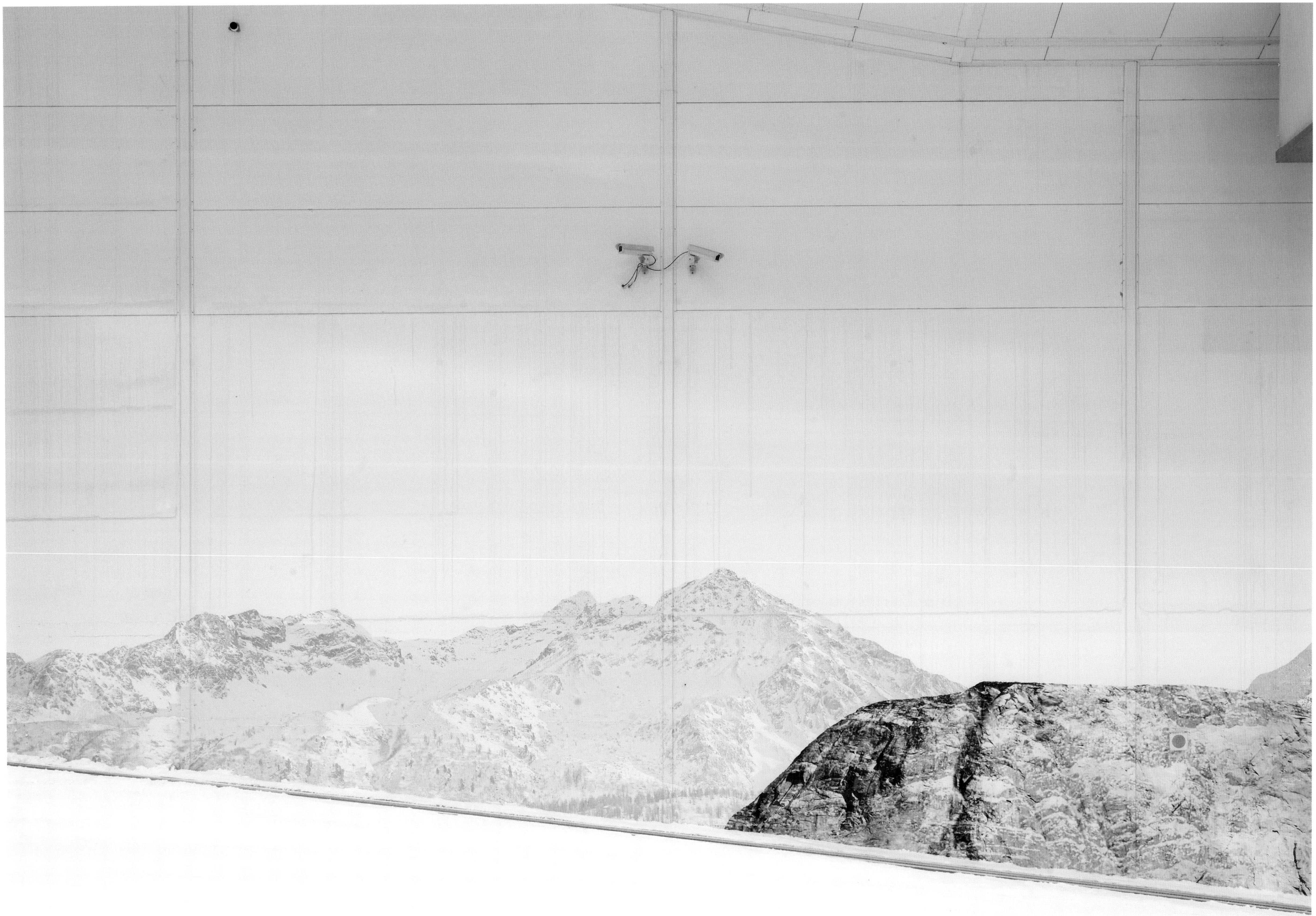

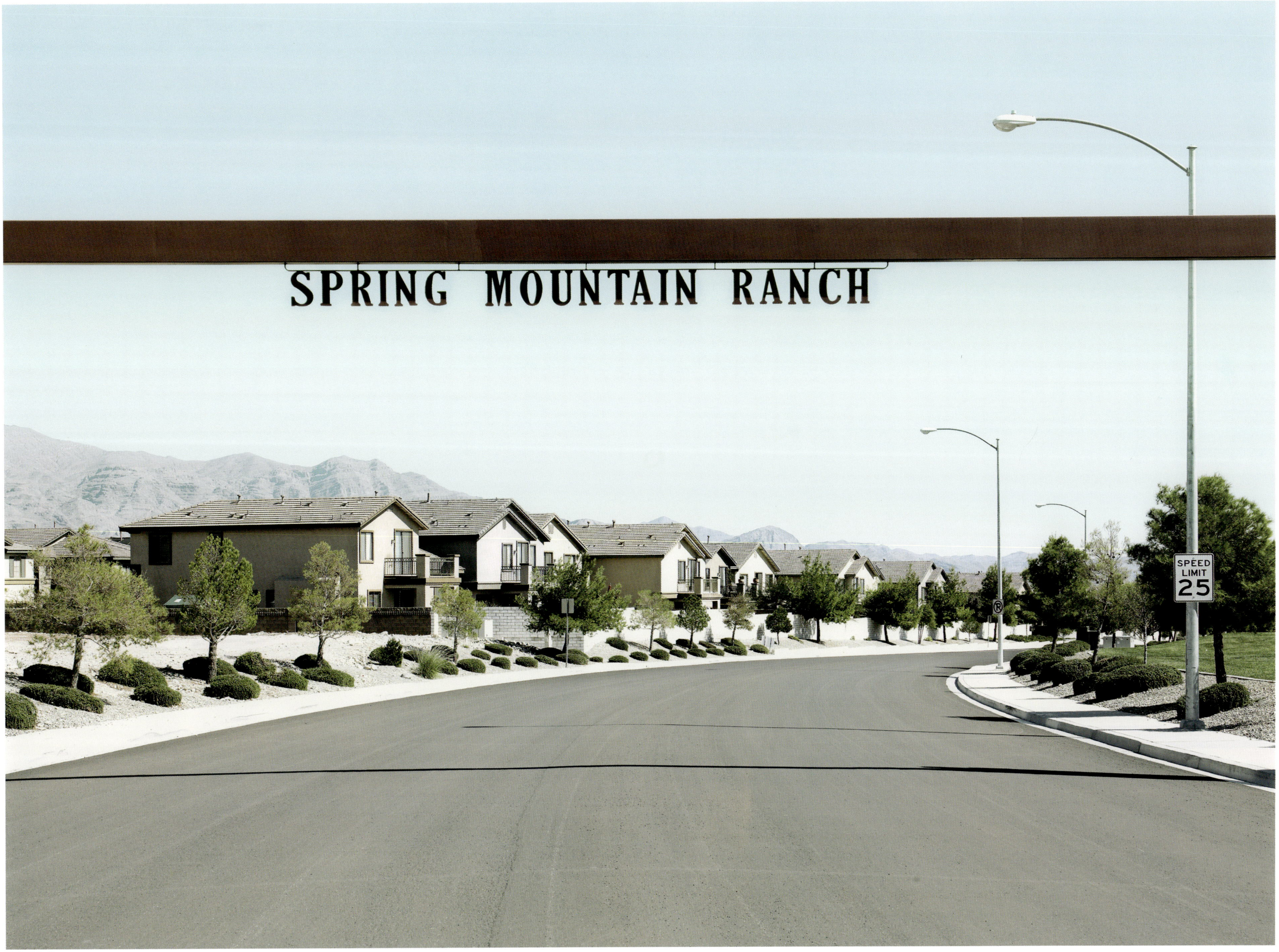
SPRING MOUNTAIN RANCH
SPEED
LIMIT
25

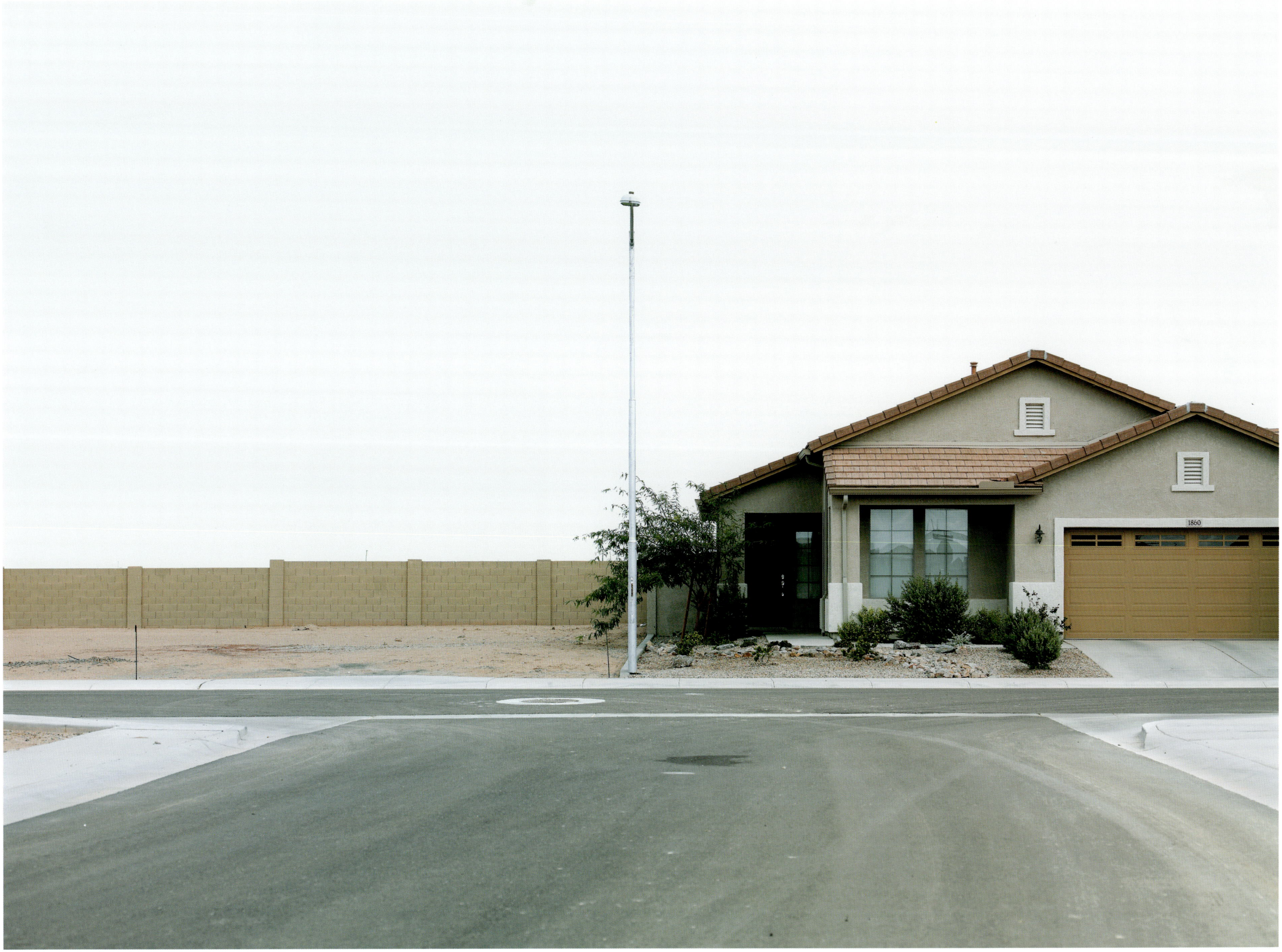
1860

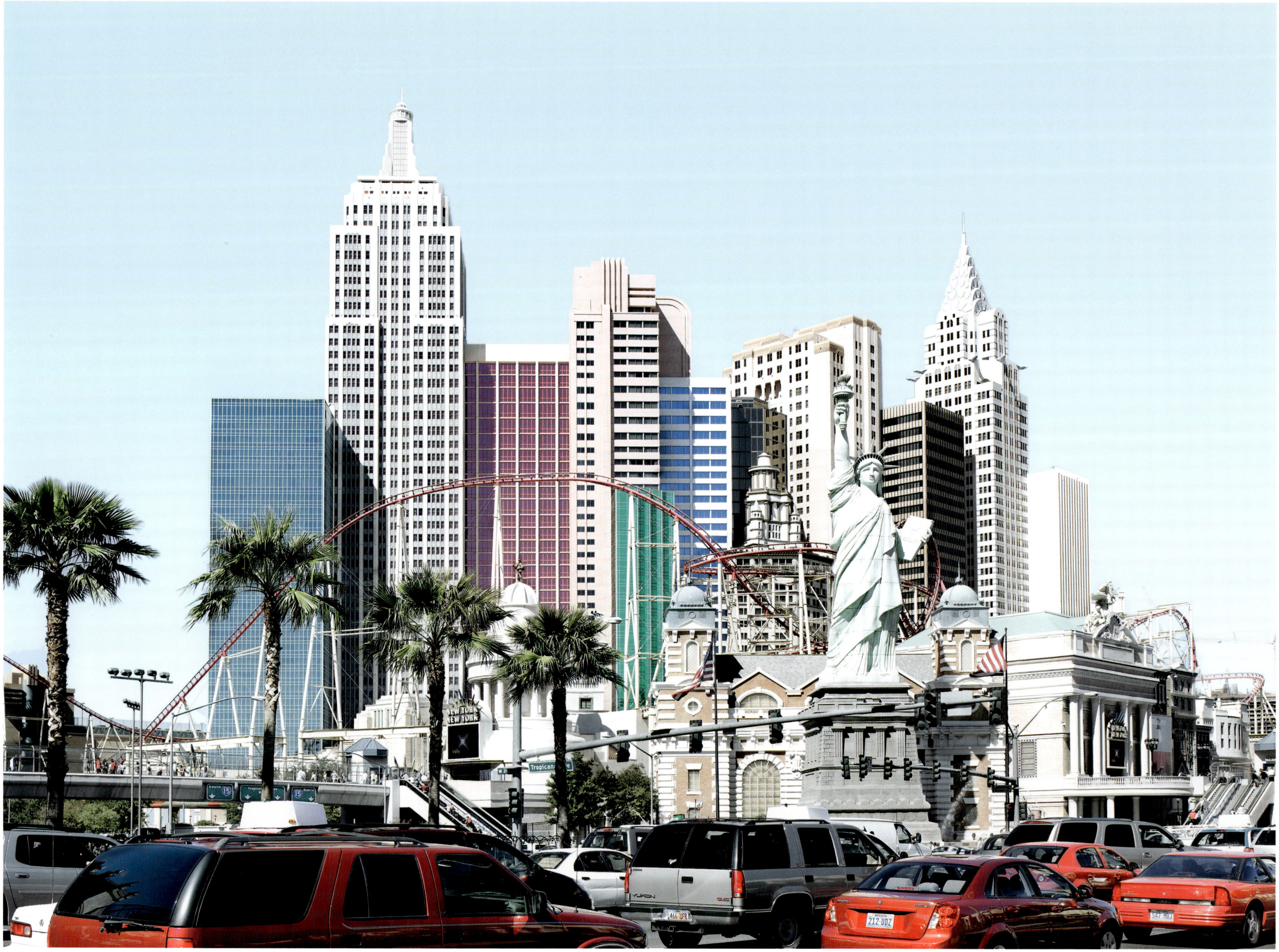
NEW YORK
NEW YORK
Tropicana
YUKON
212 UDZ

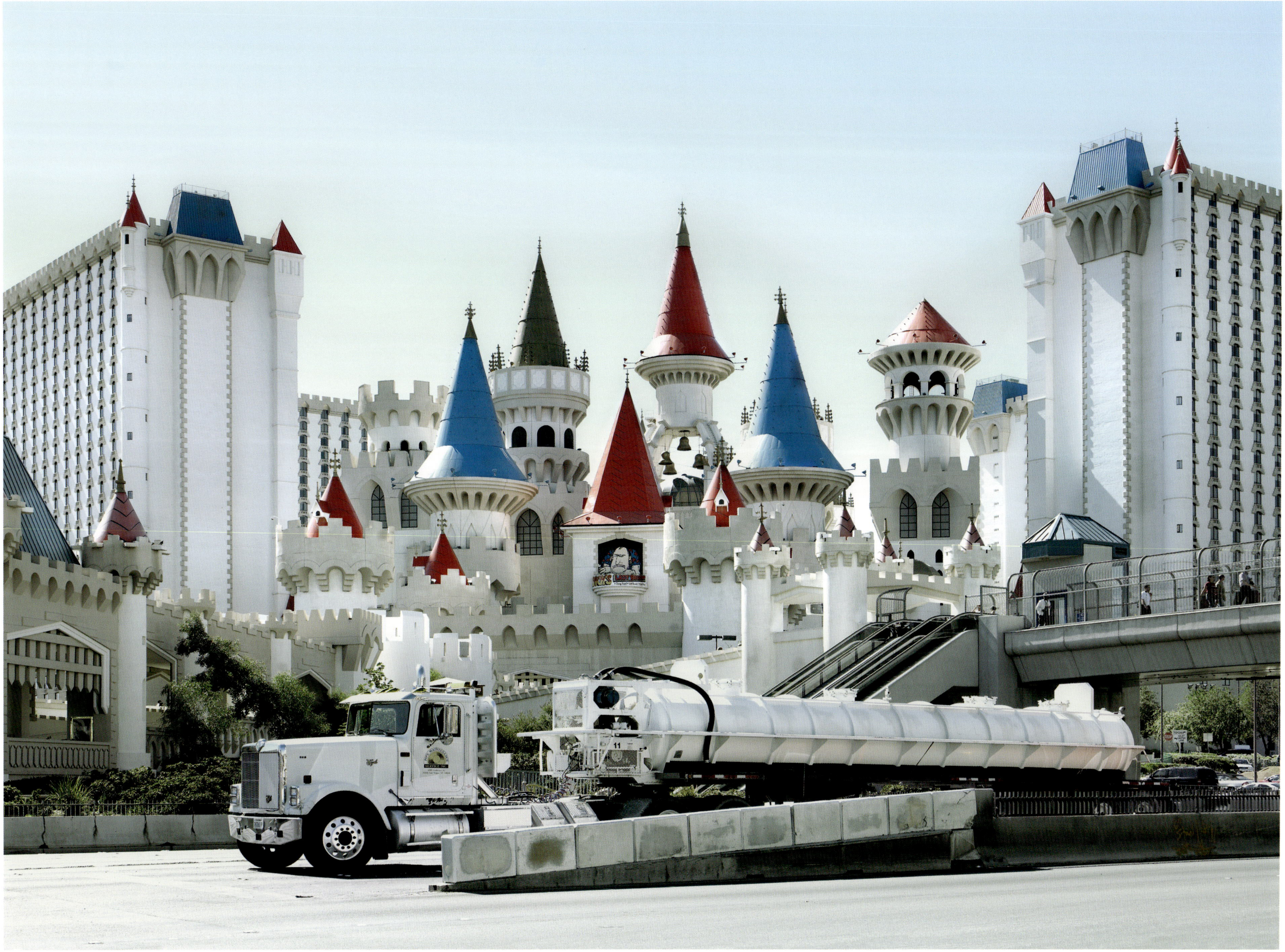

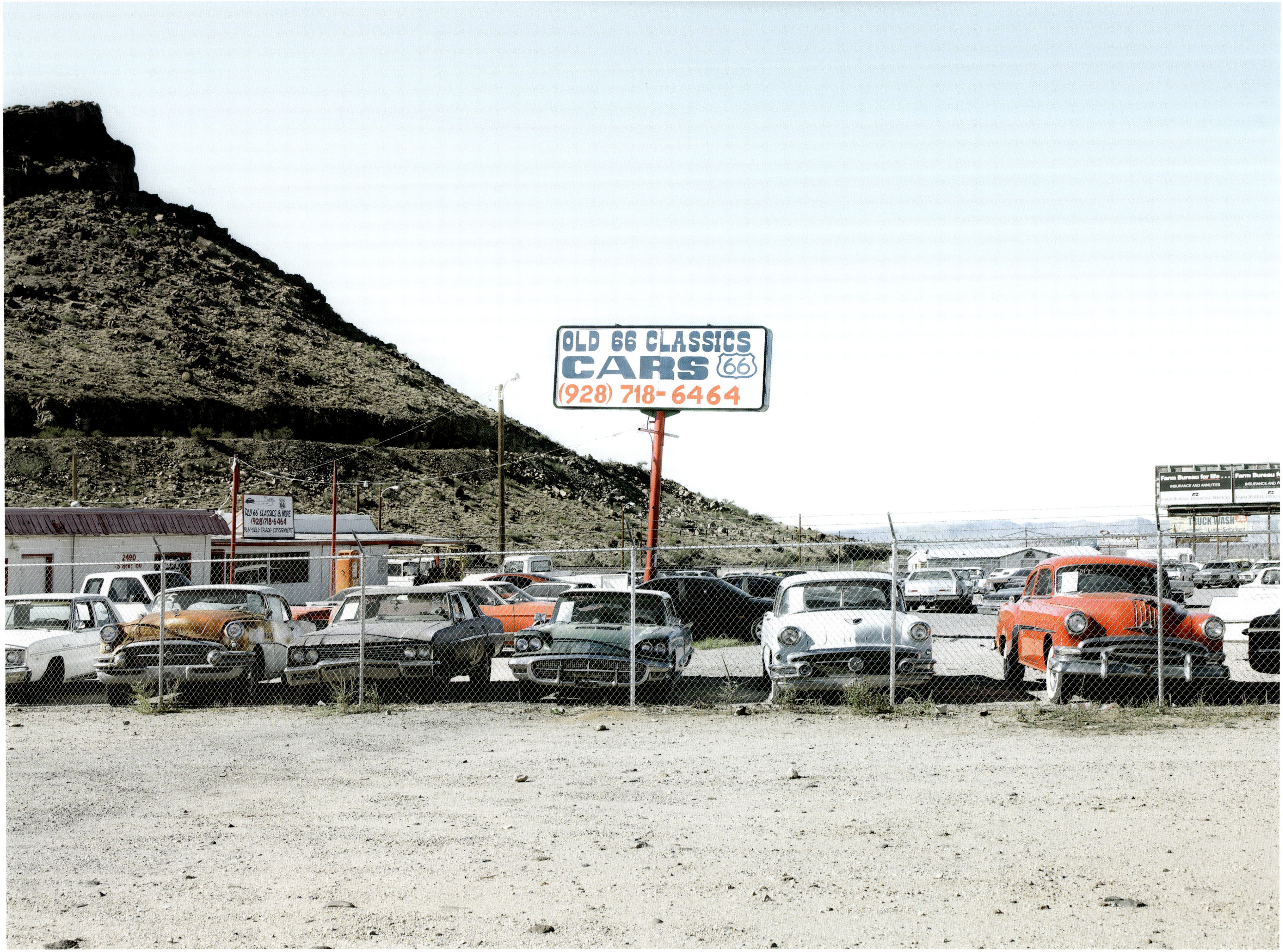
OLD 66 CLASSICS
CARS
66
(928) 718-6464
OLD 66 CLASSICS & MORE
(928) 718-6464

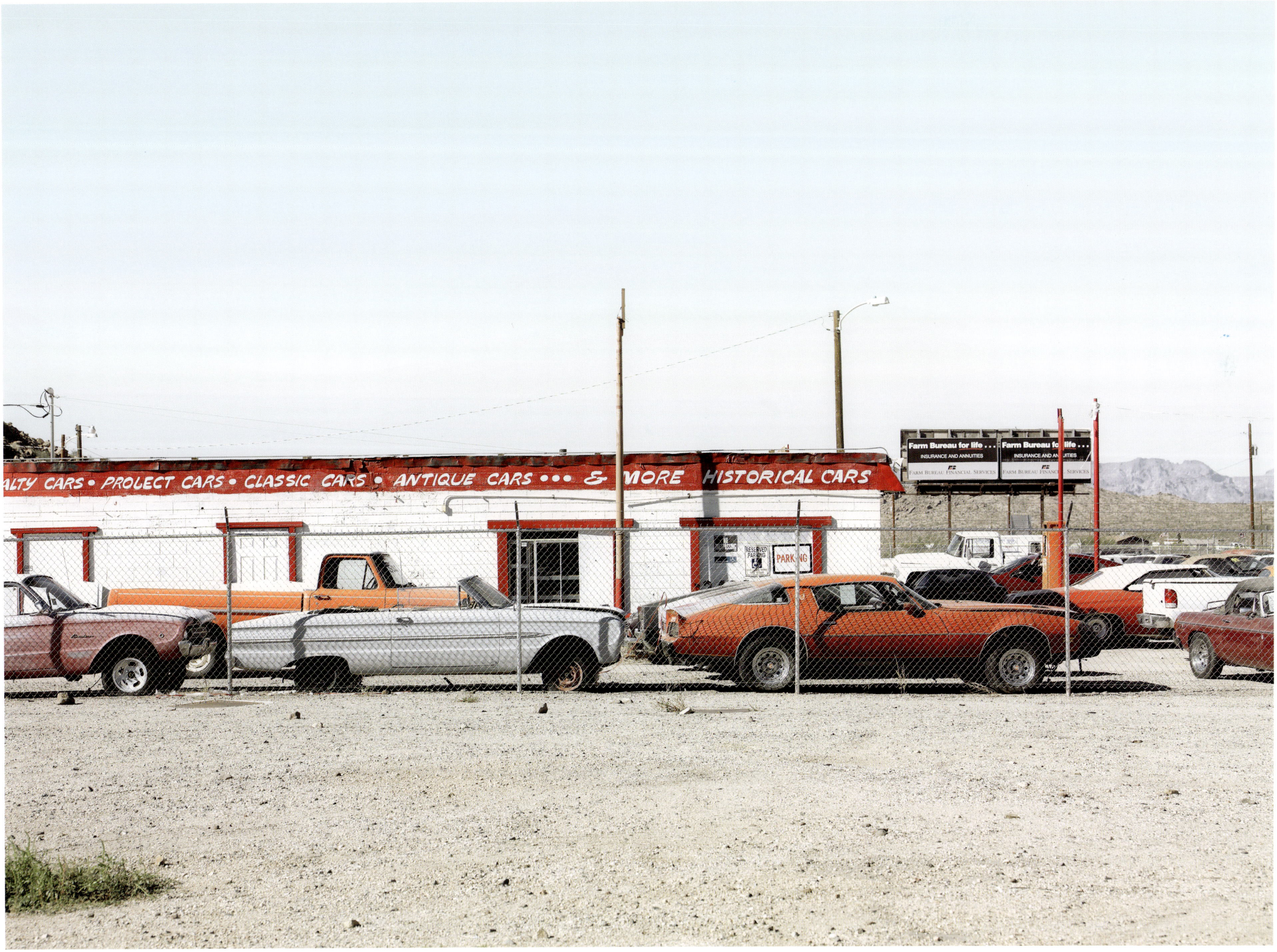
ALTY CARS • PROLECT CARS • CLASSIC CARS • ANTIQUE CARS ••• & MORE HISTORICAL CARS
Farm Bureau for life . . .
INSURANCE AND ANNUITIES
FARM BUREAU FINANCIAL SERVICES
RESERVED PARKING

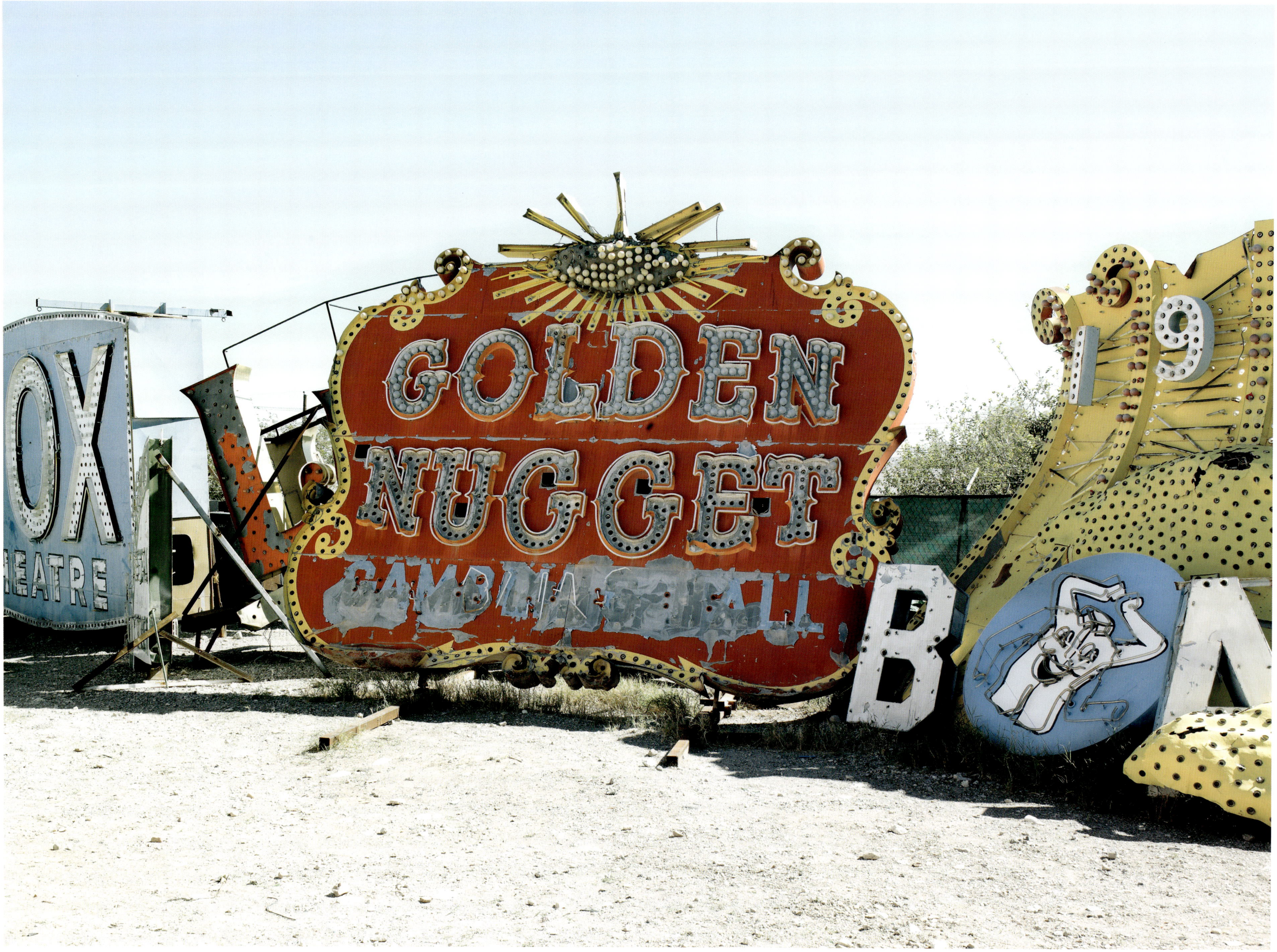
OX
HEATRE
GOLDEN
NUGGET
19
B

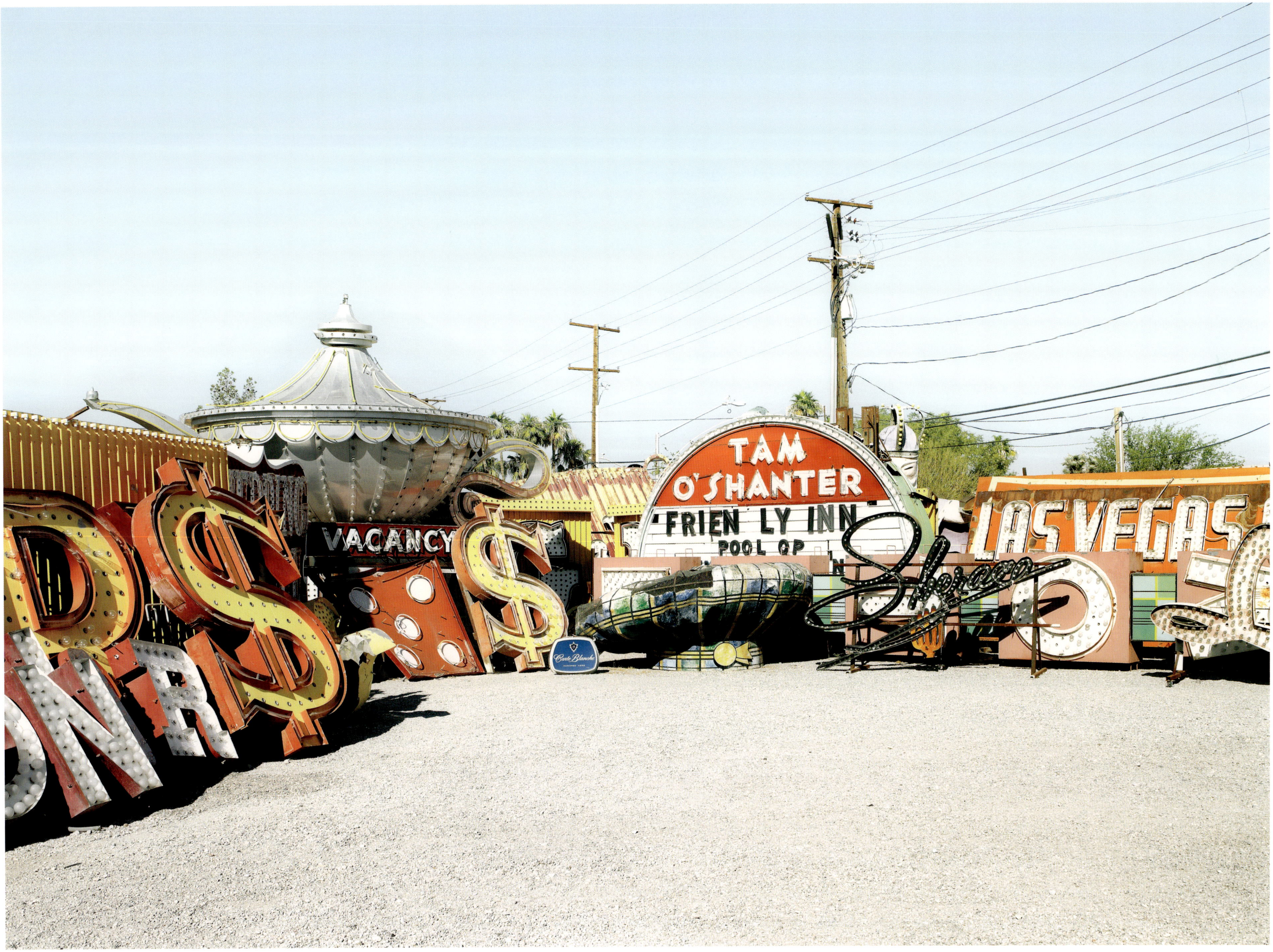
TAM
O'SHANTER
FRIEN LY INN
POOL OP
VACANCY
LAS VEGAS

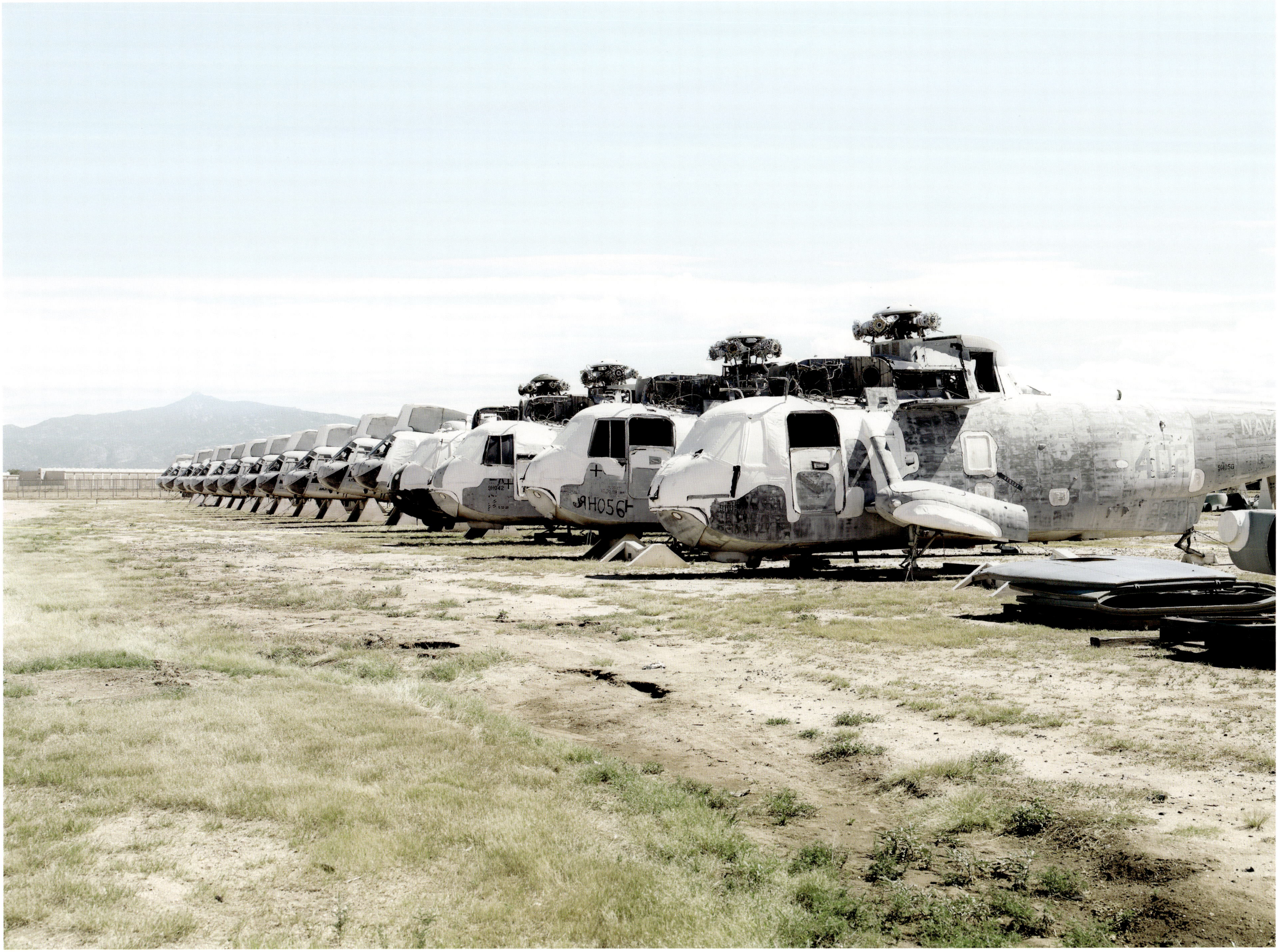

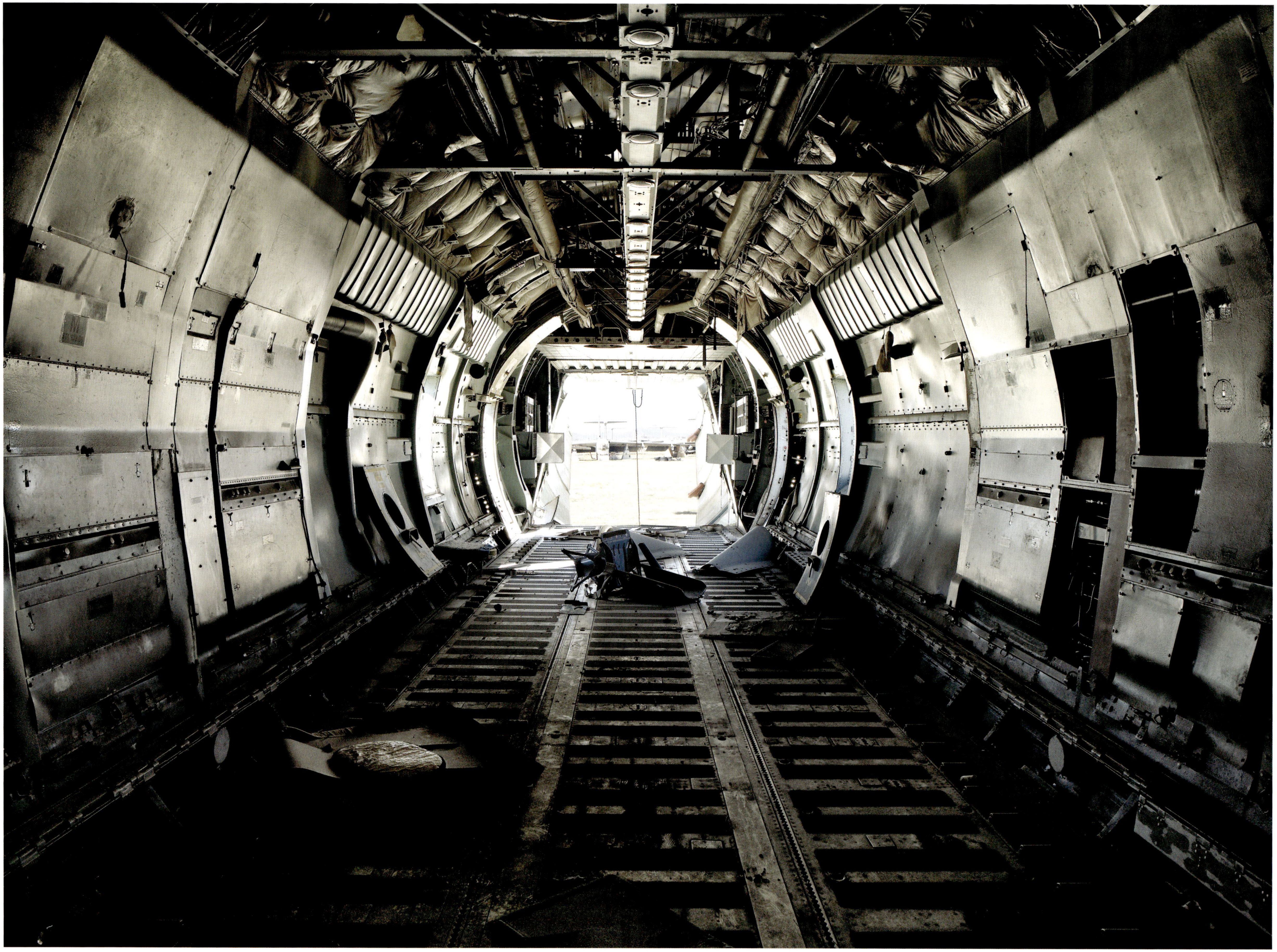

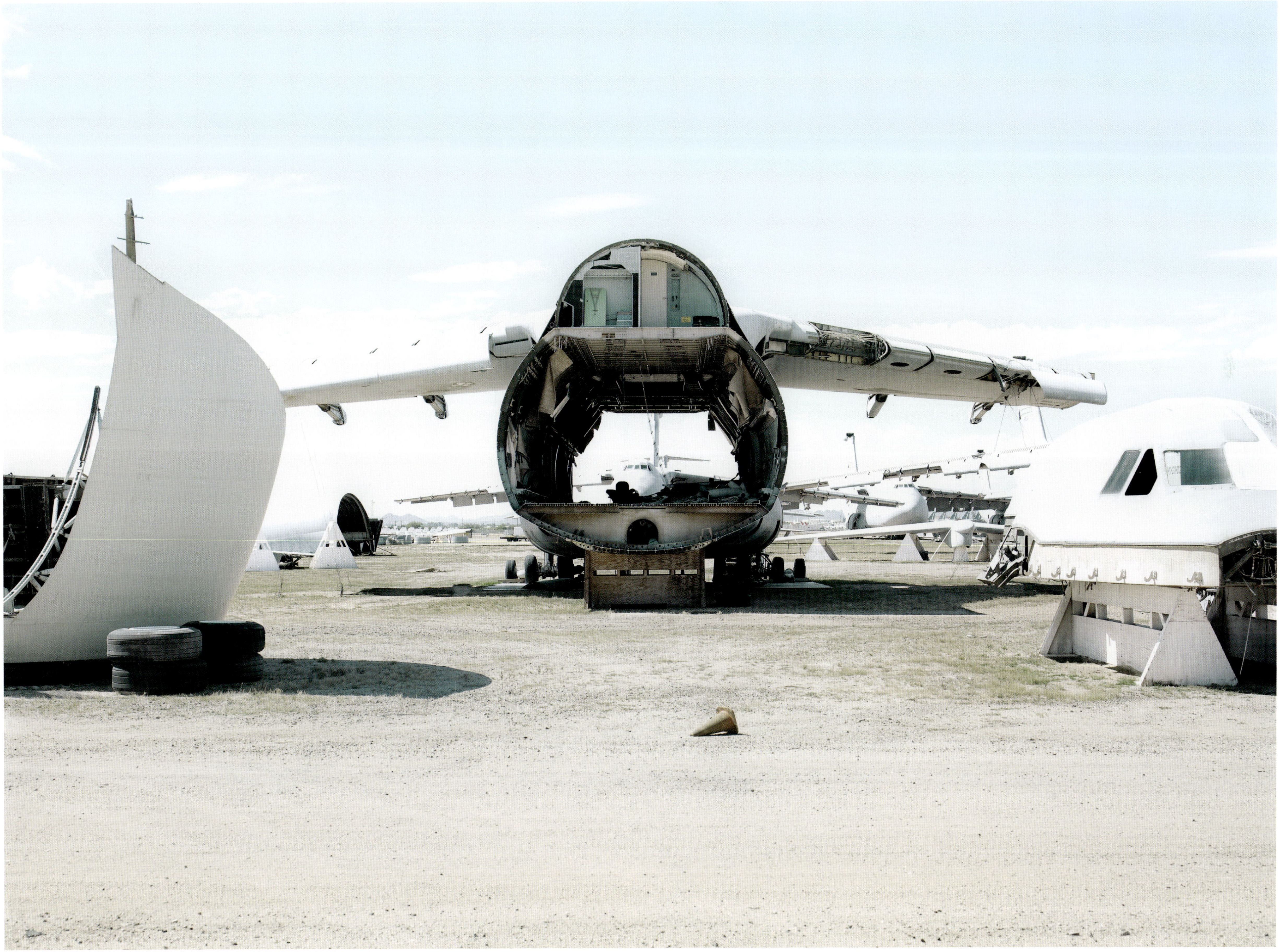

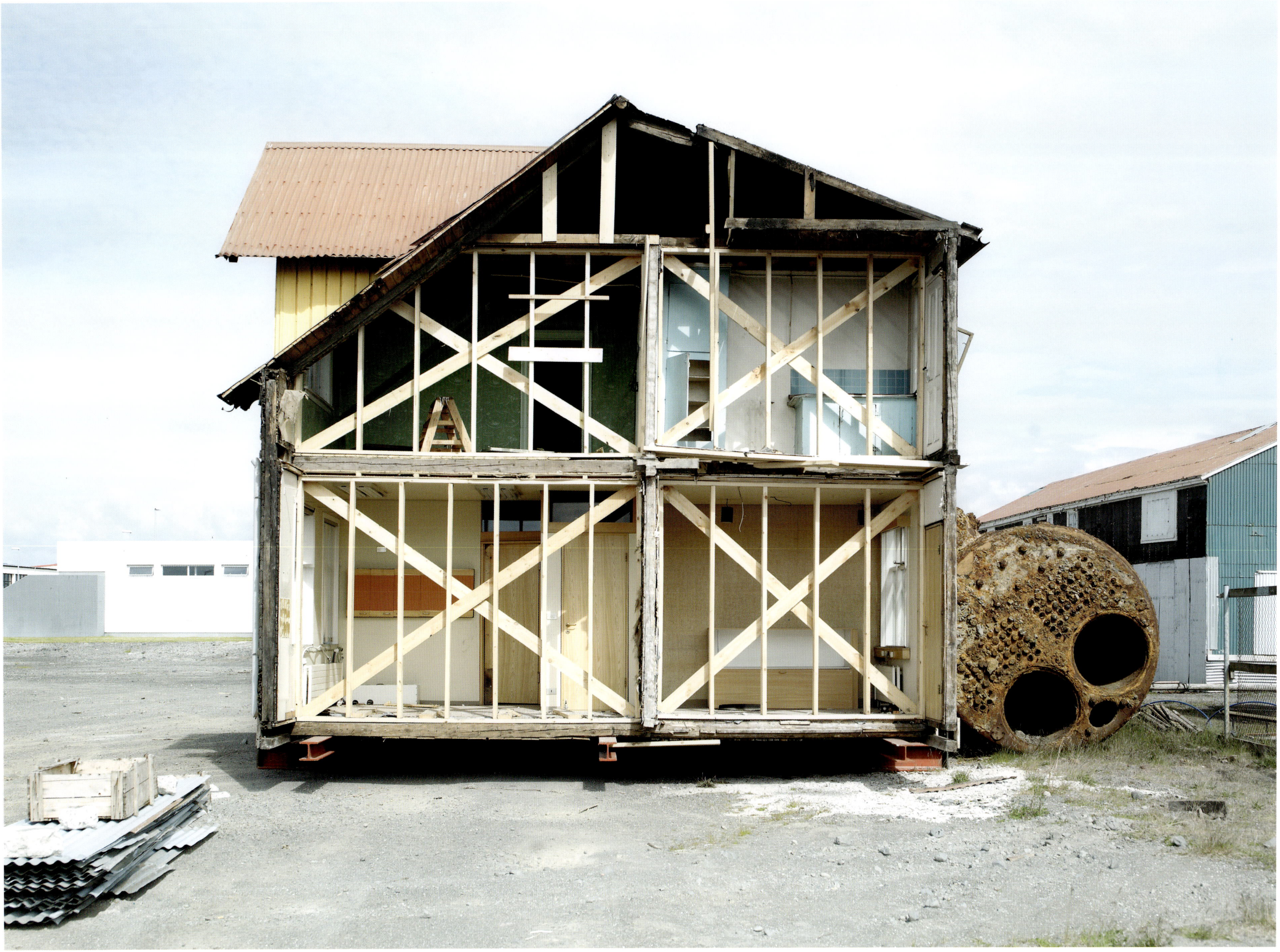

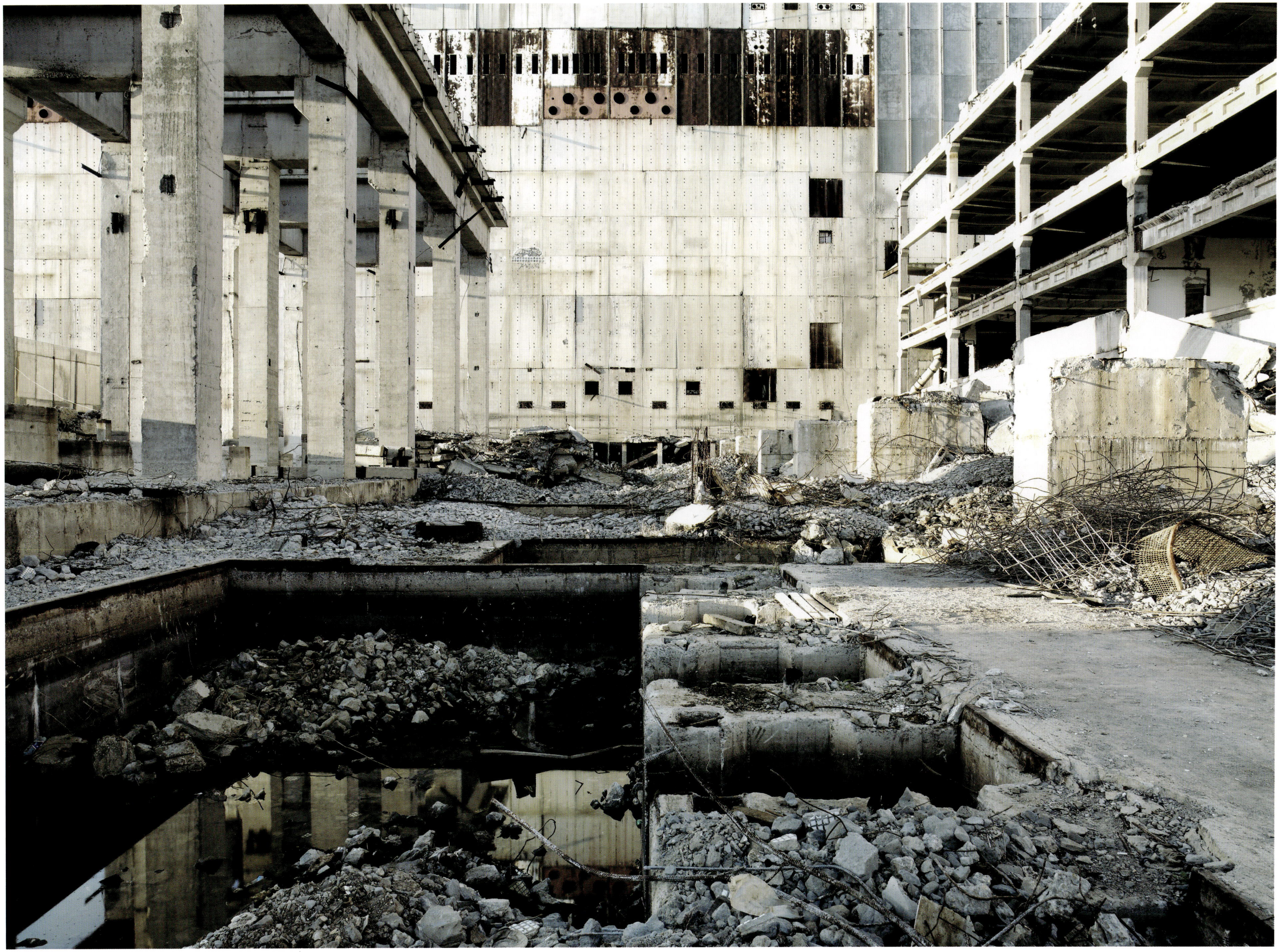

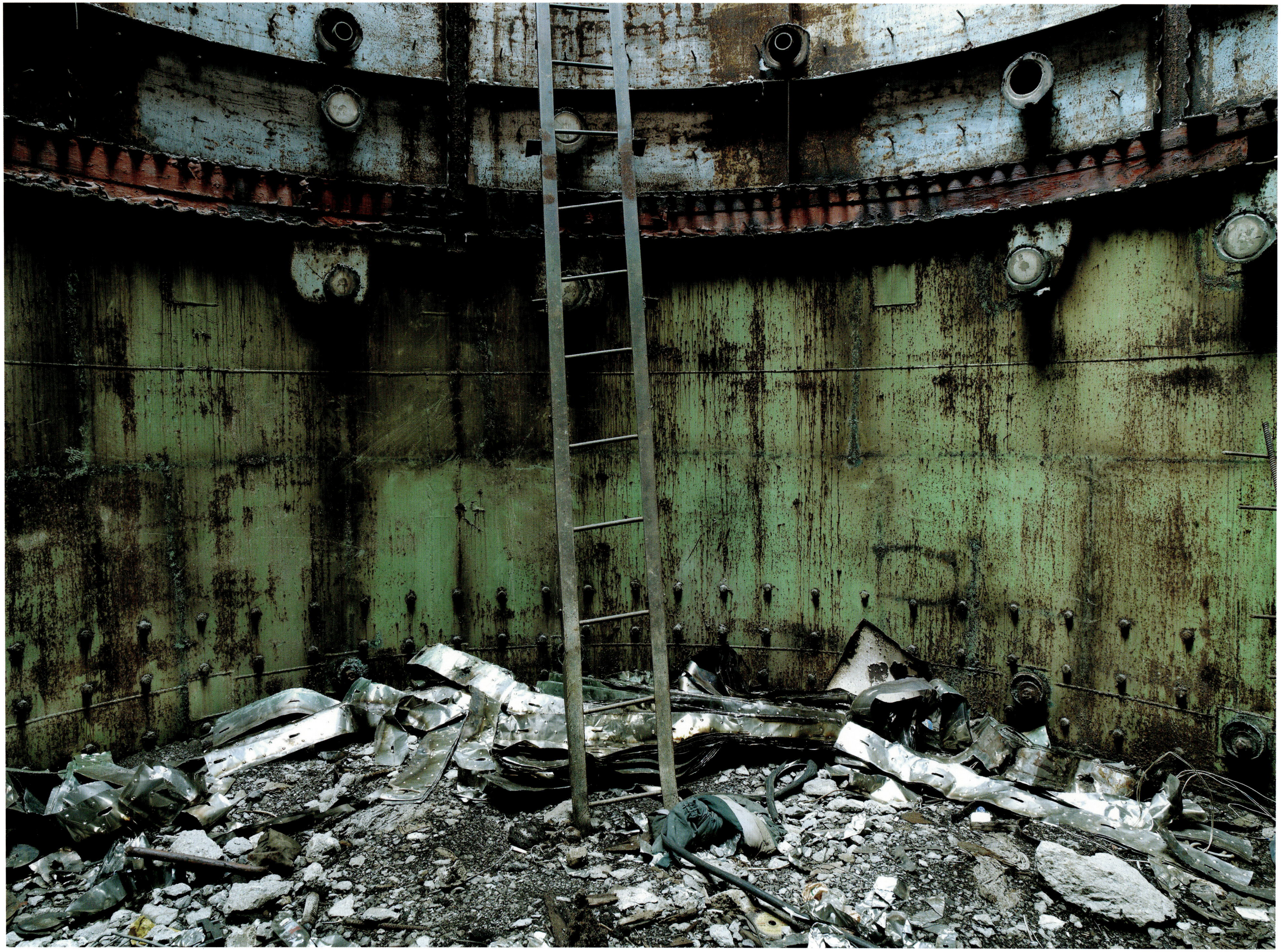

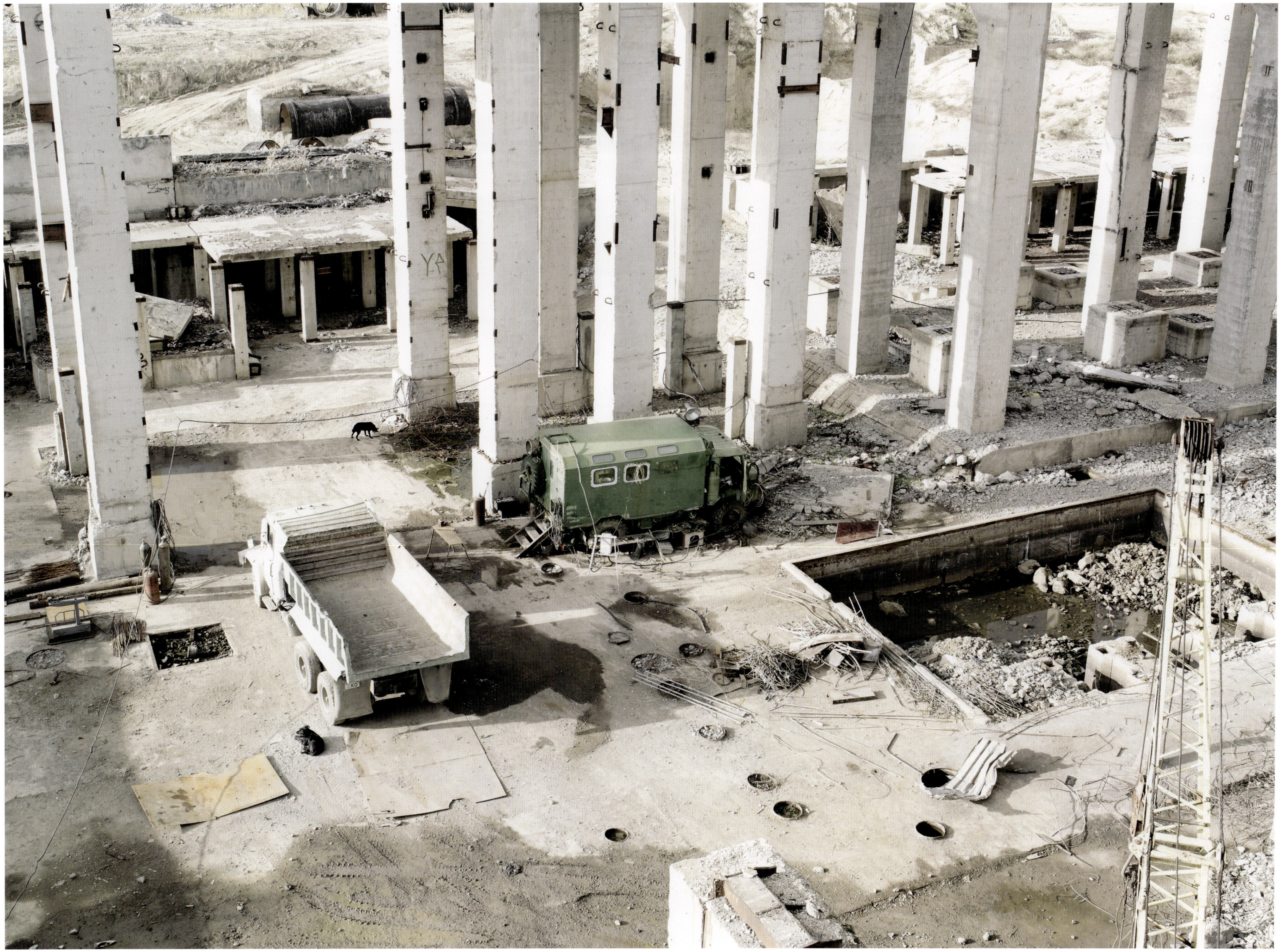

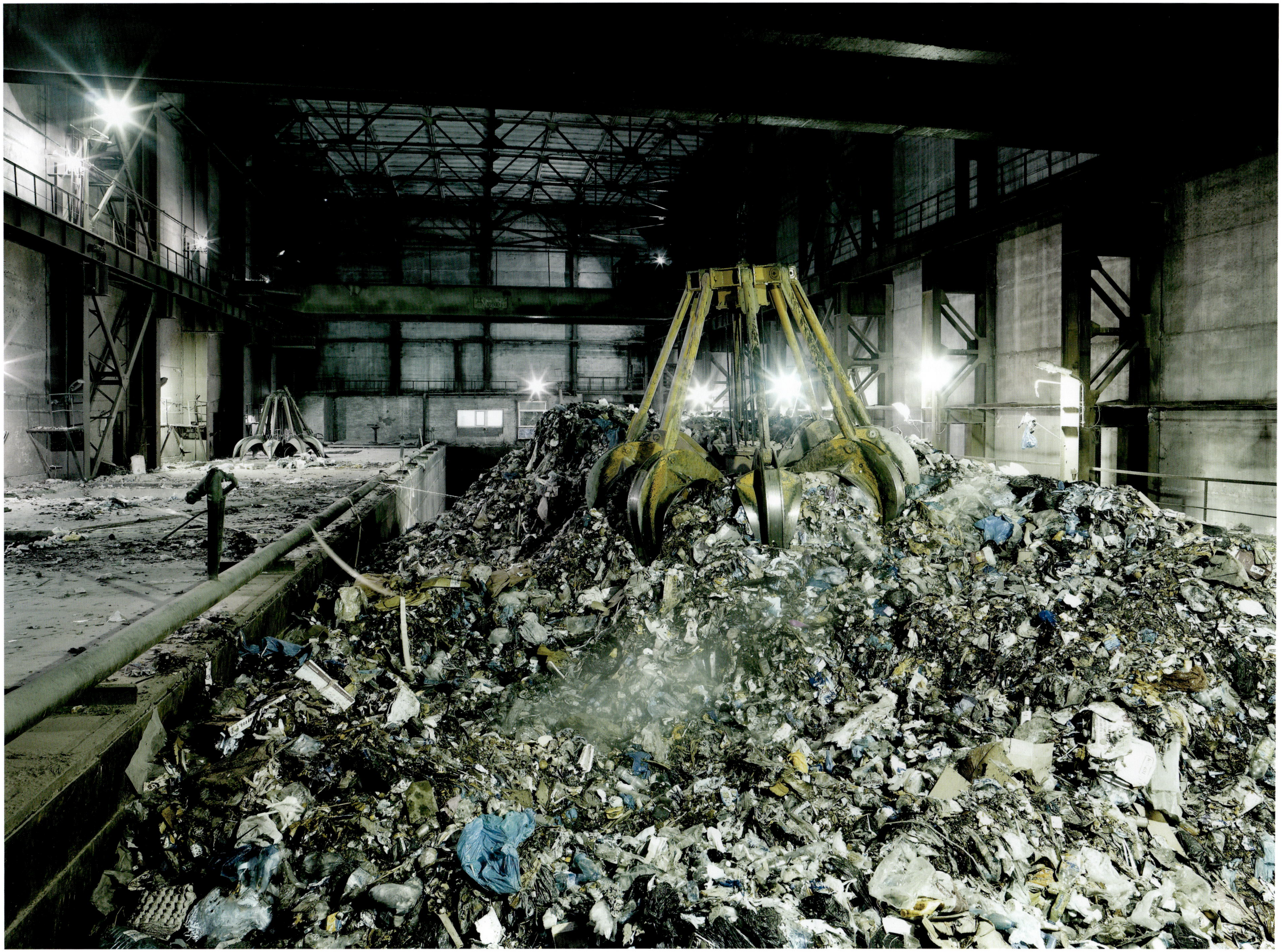

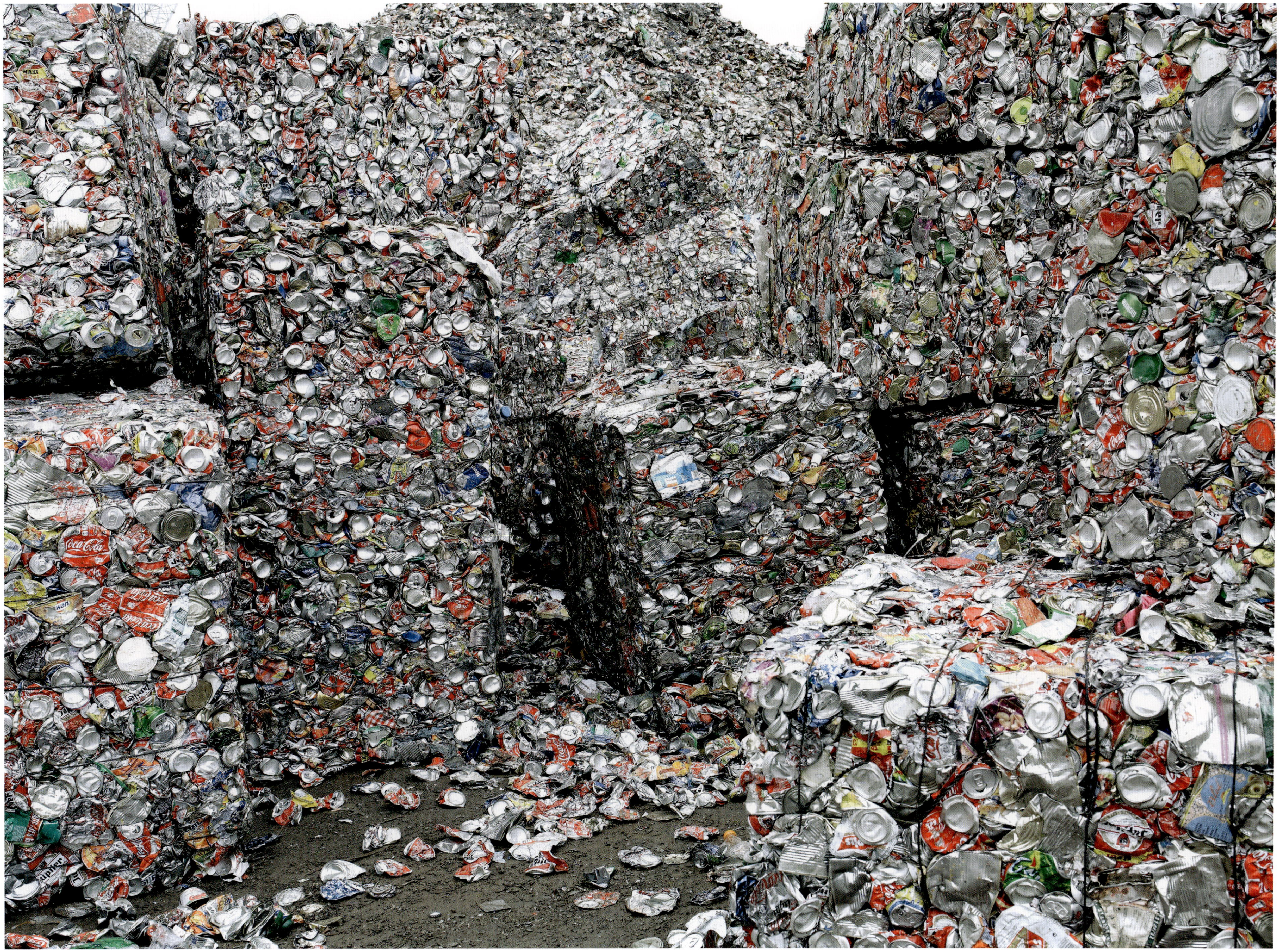

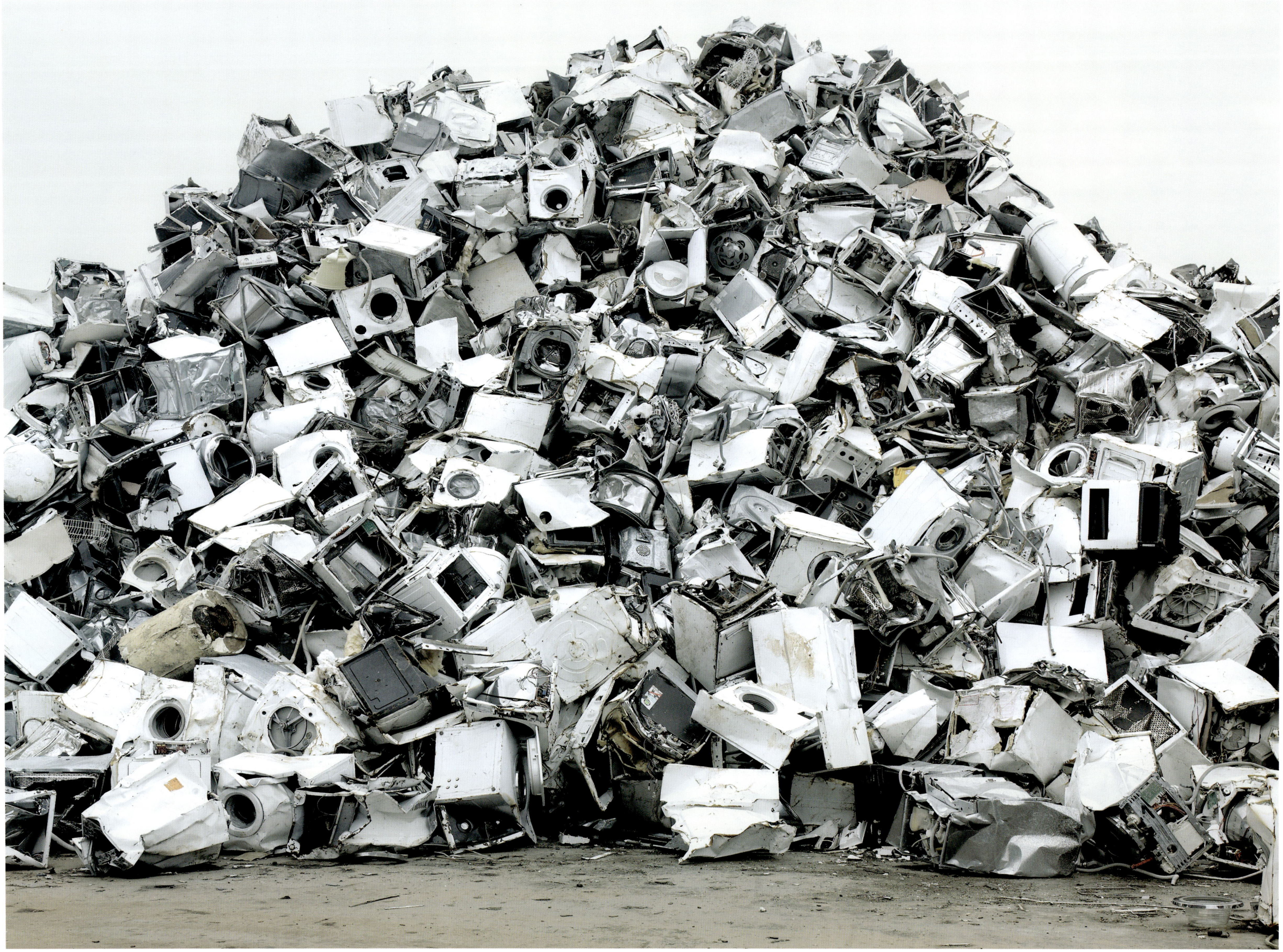

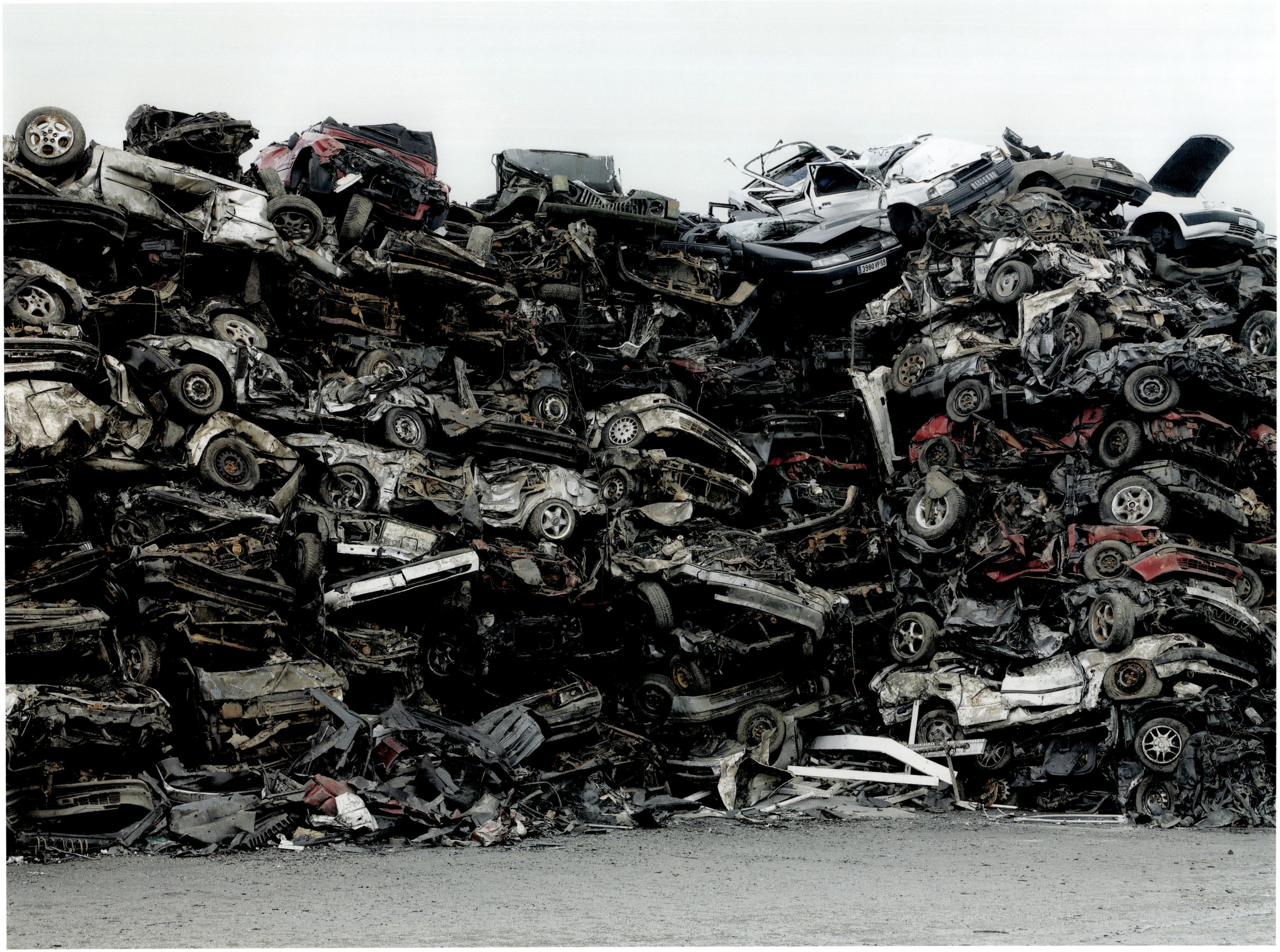

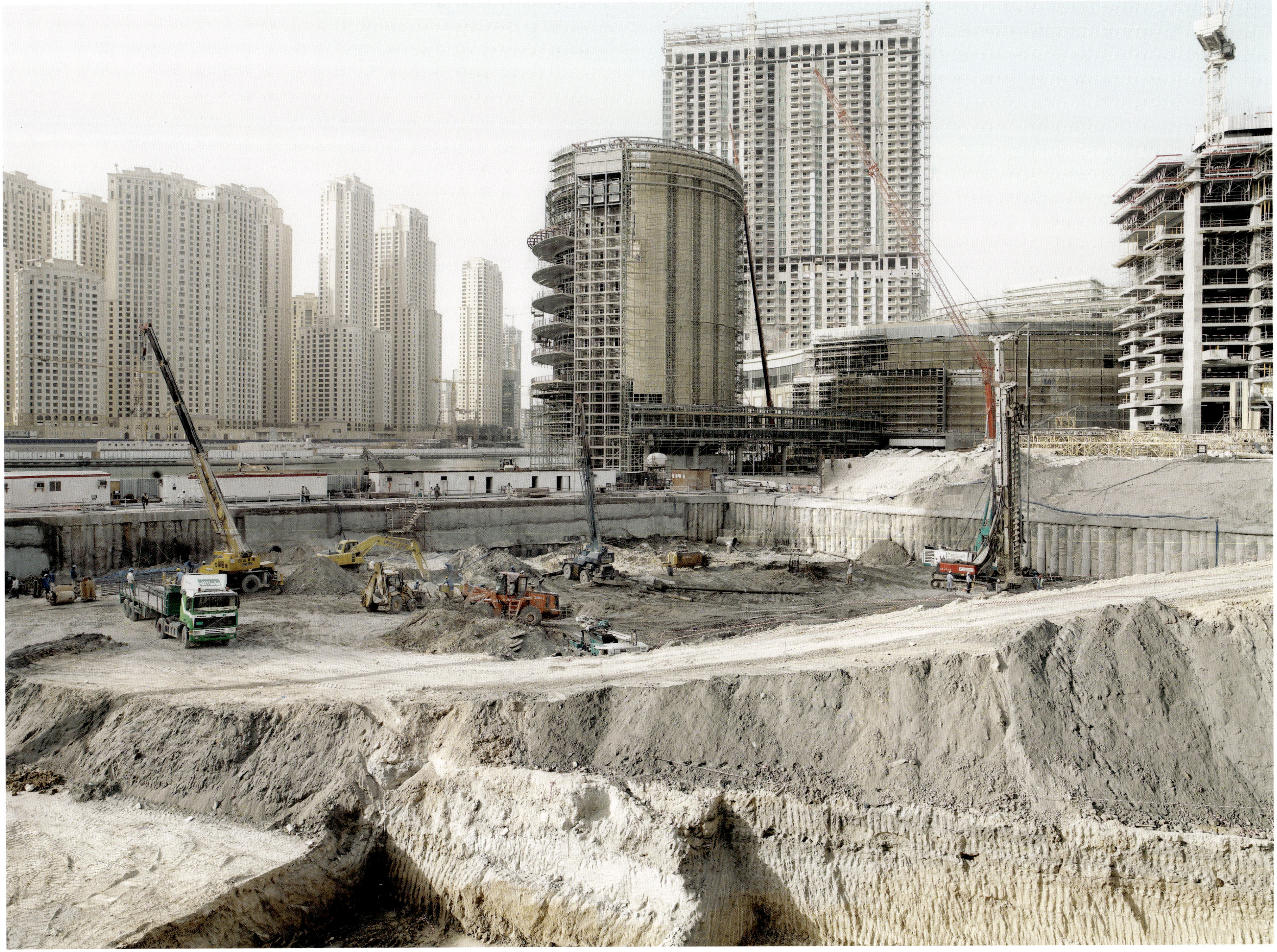

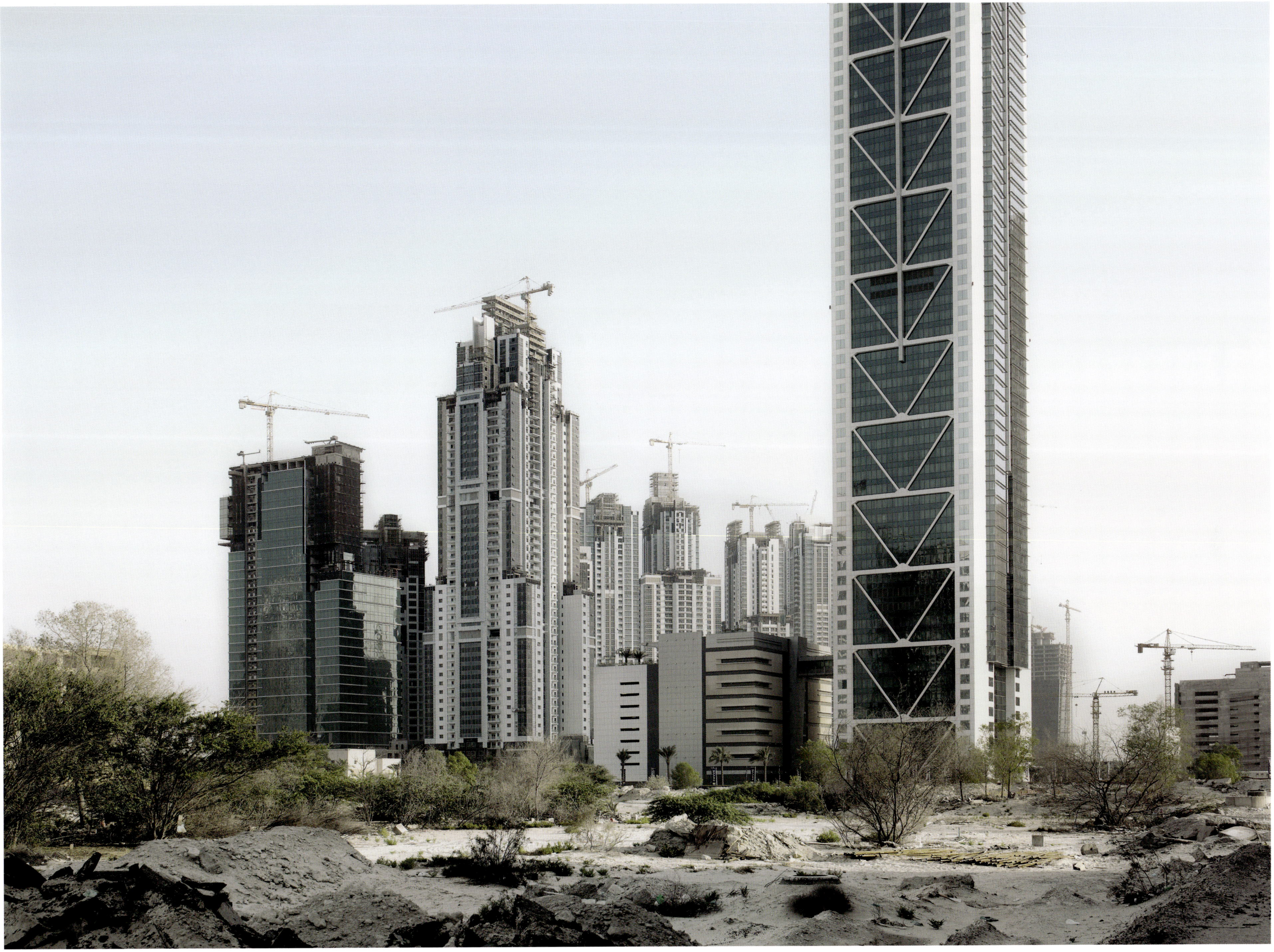

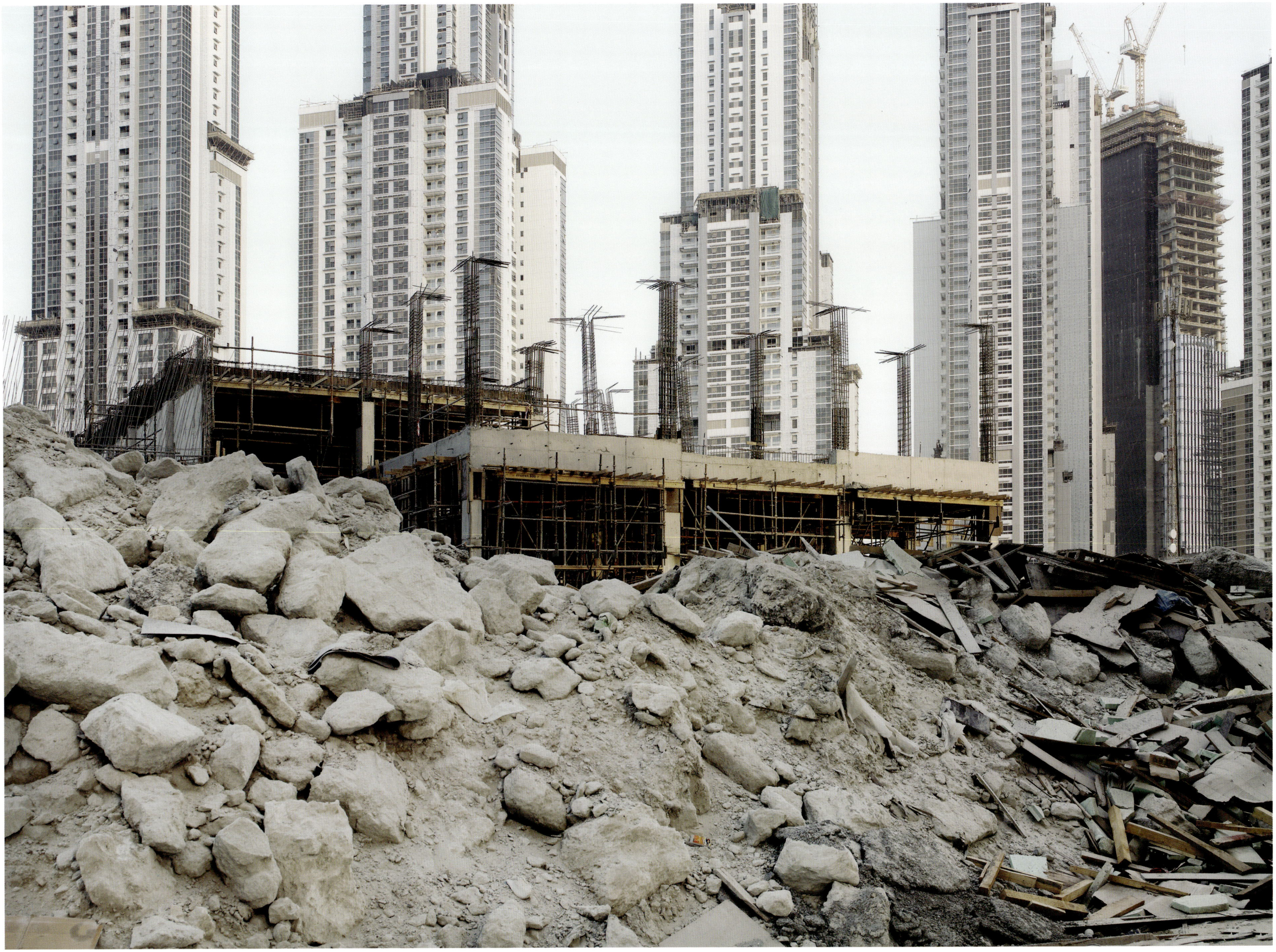

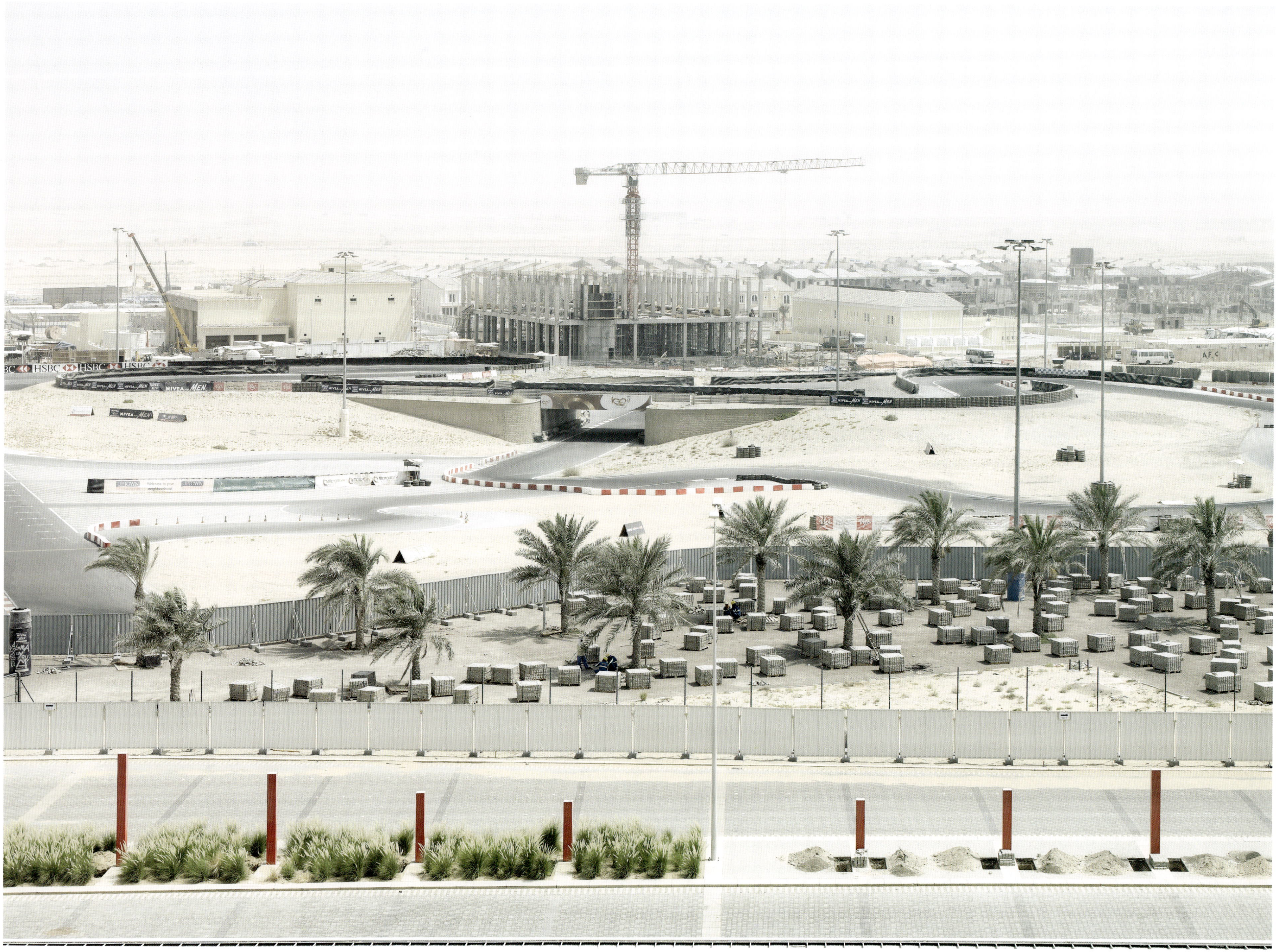
HSBC HSBC HSBC

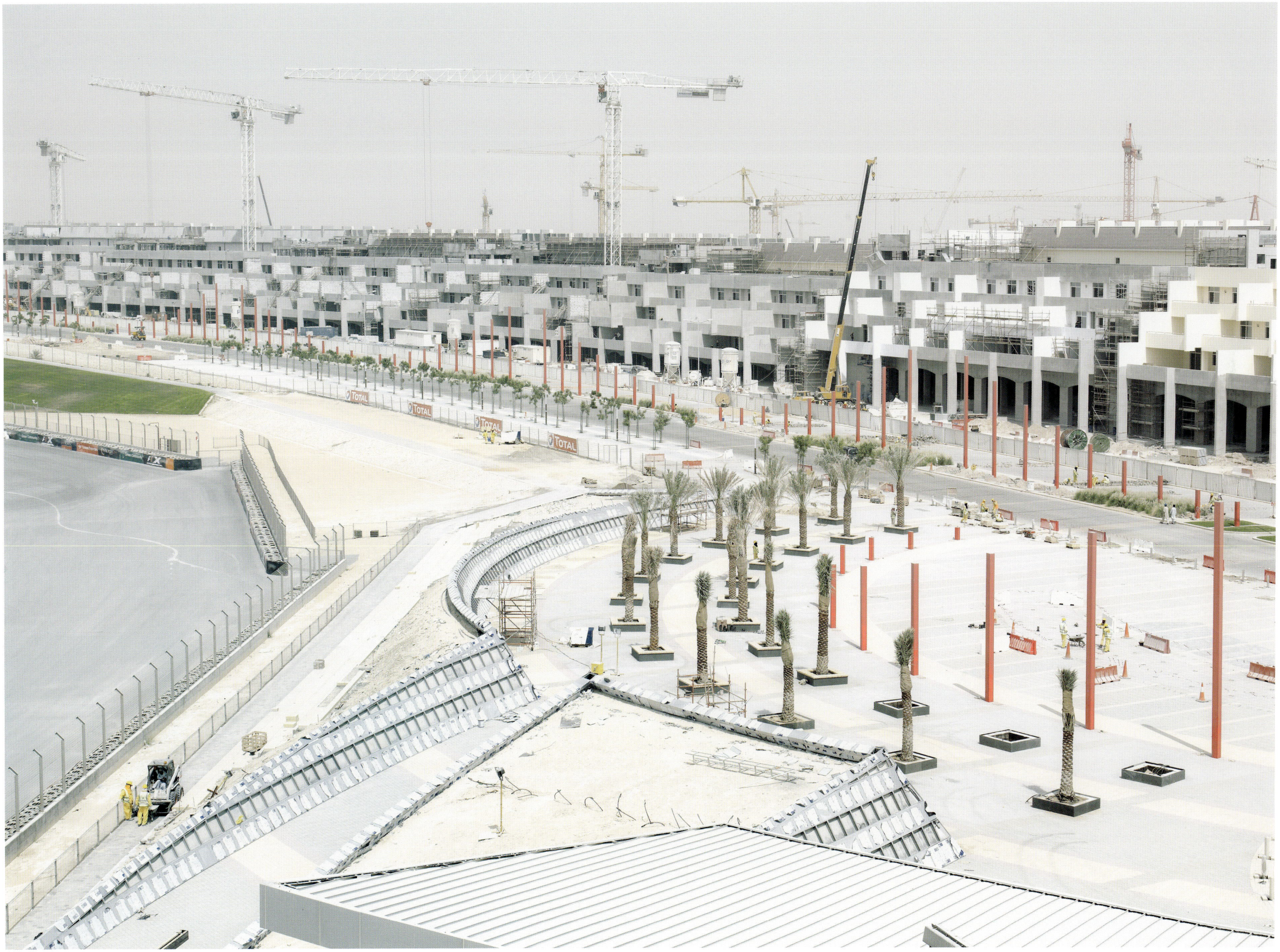
TOTAL

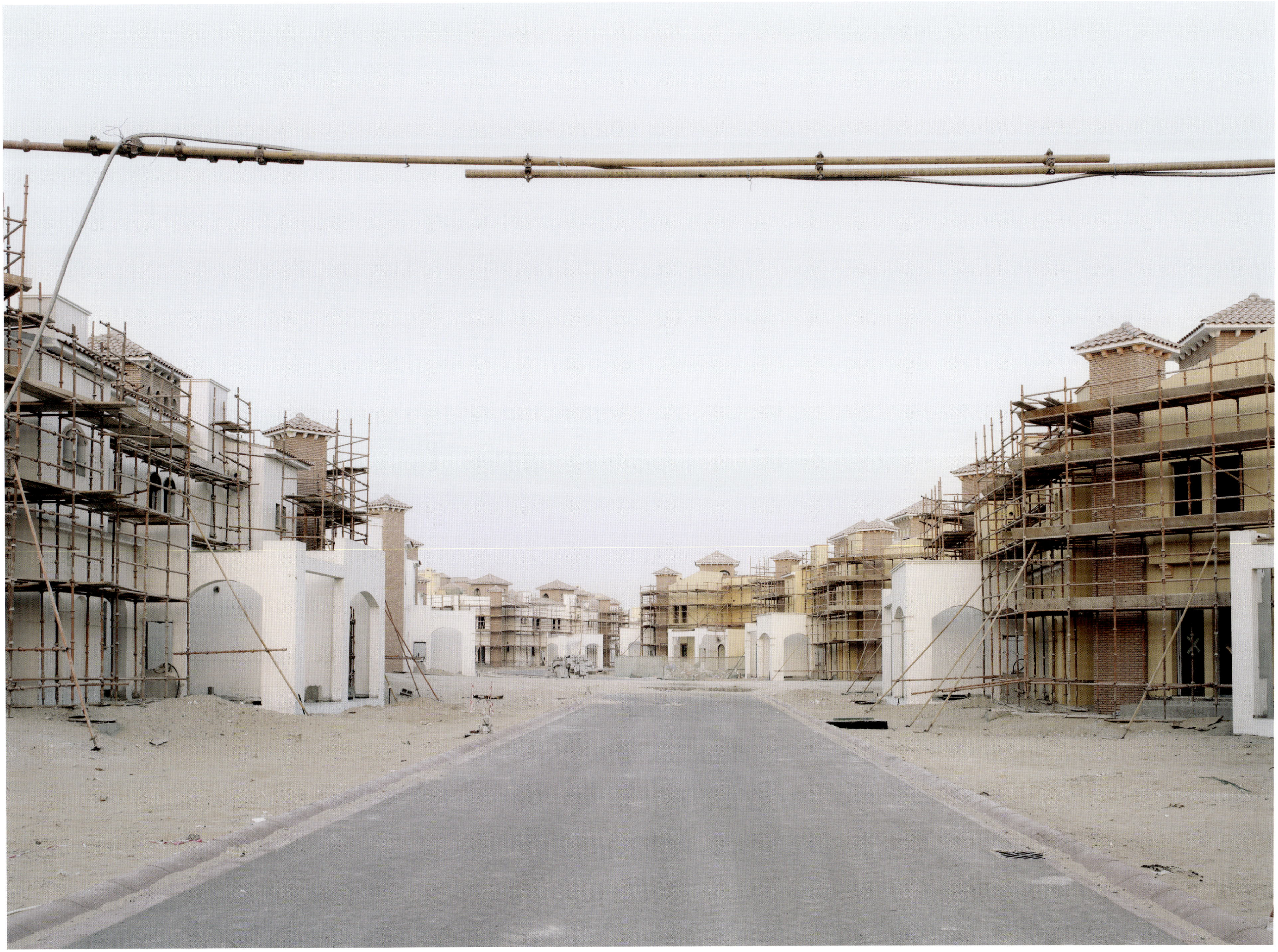

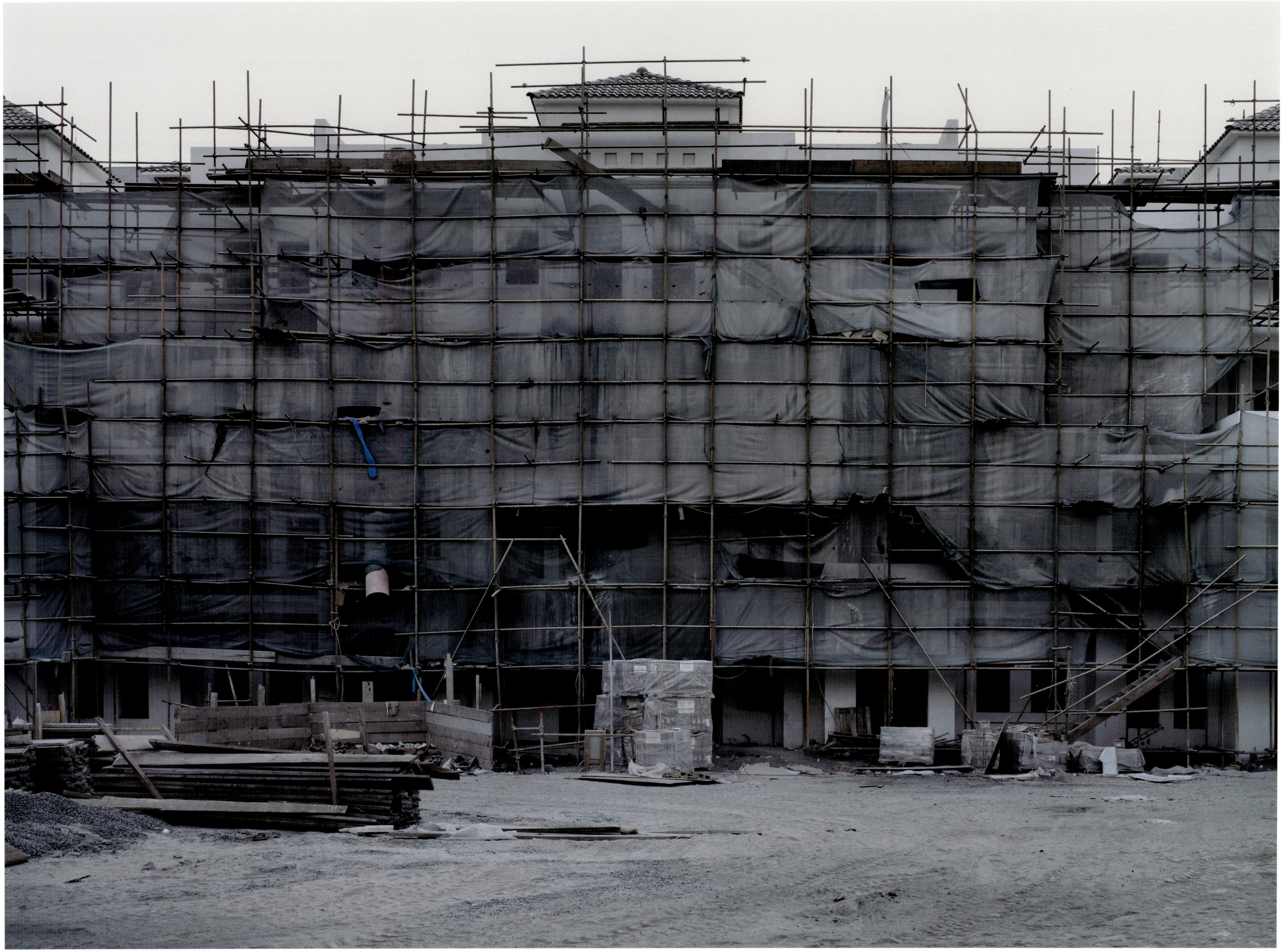

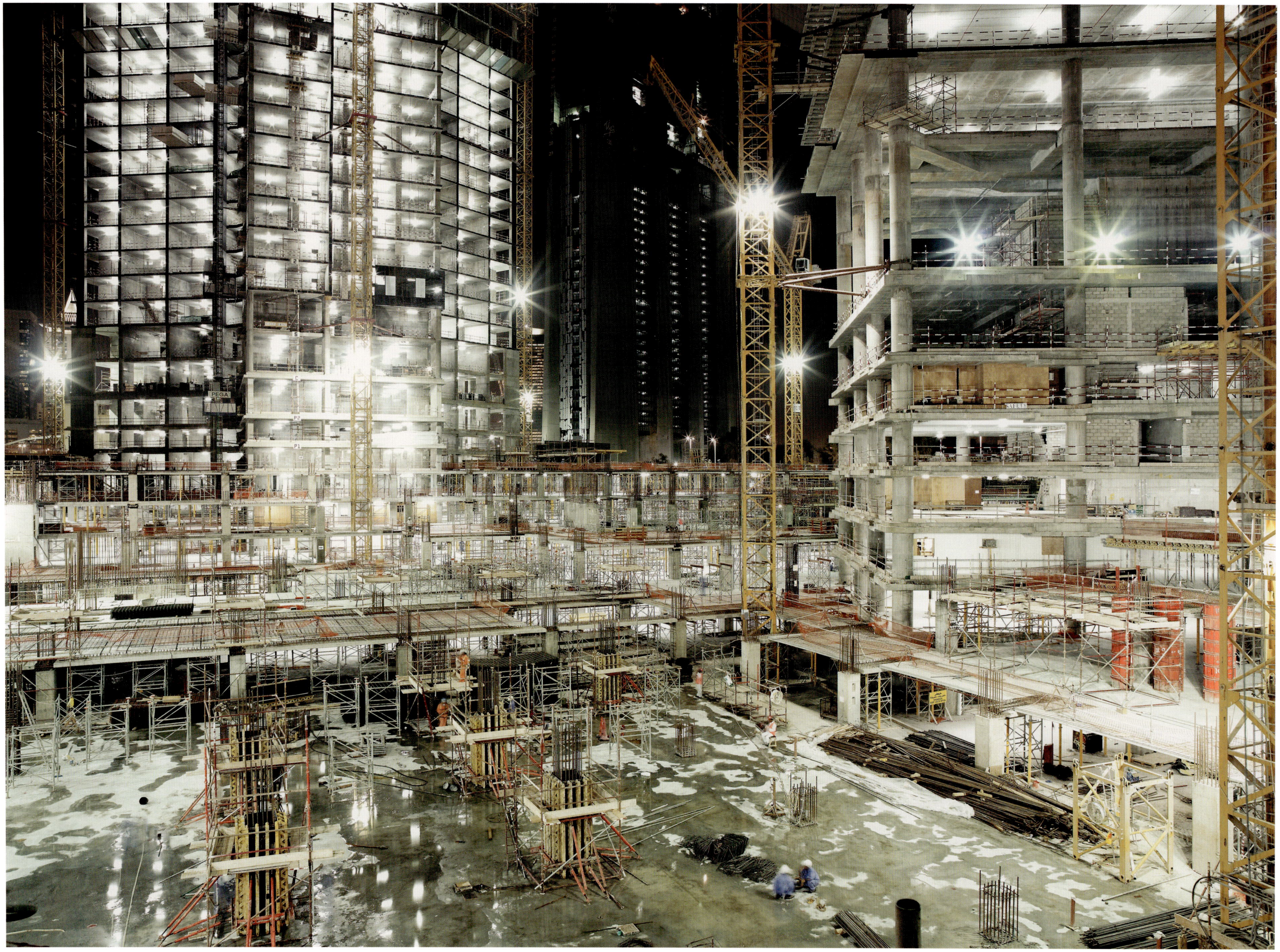

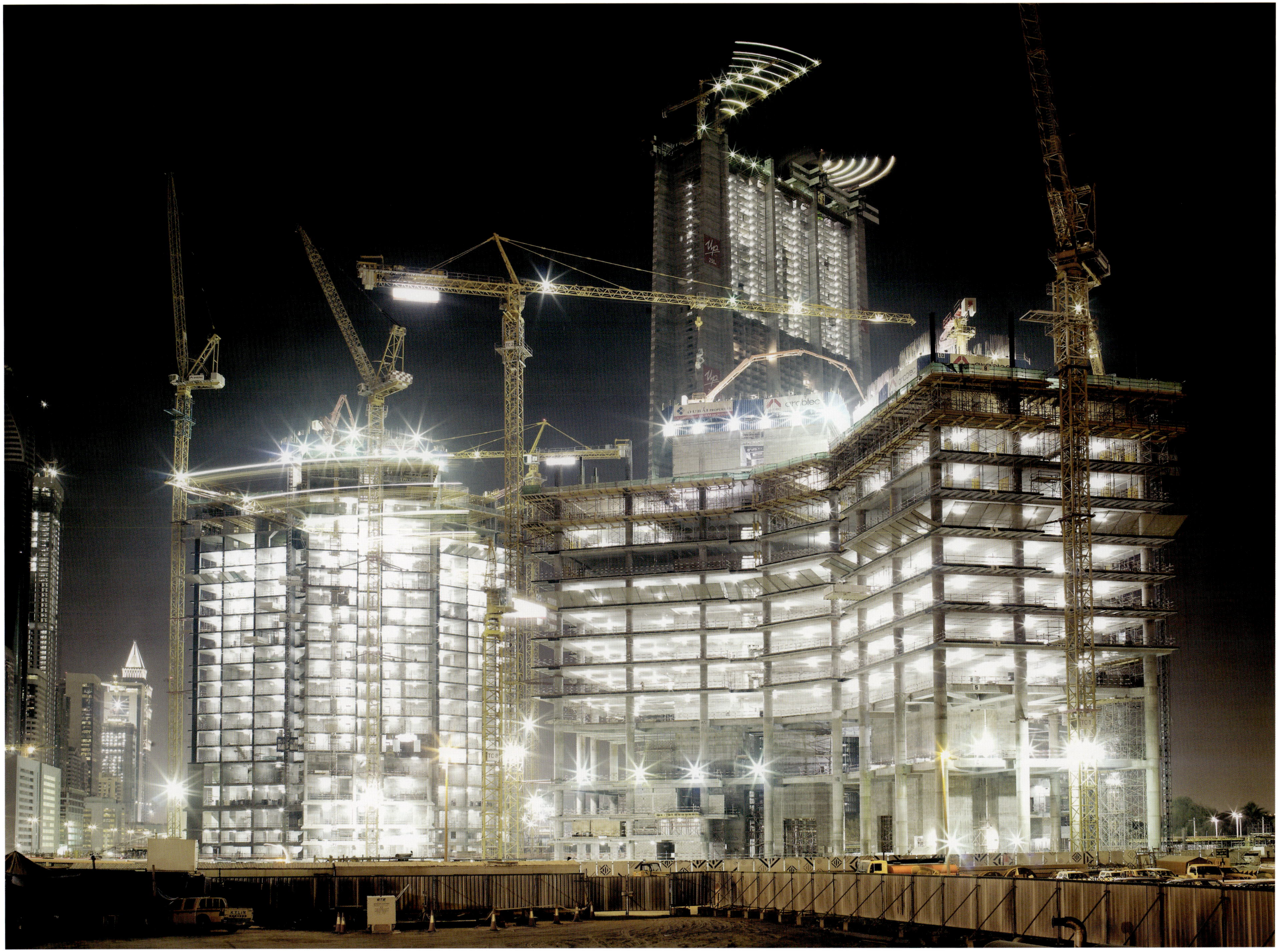

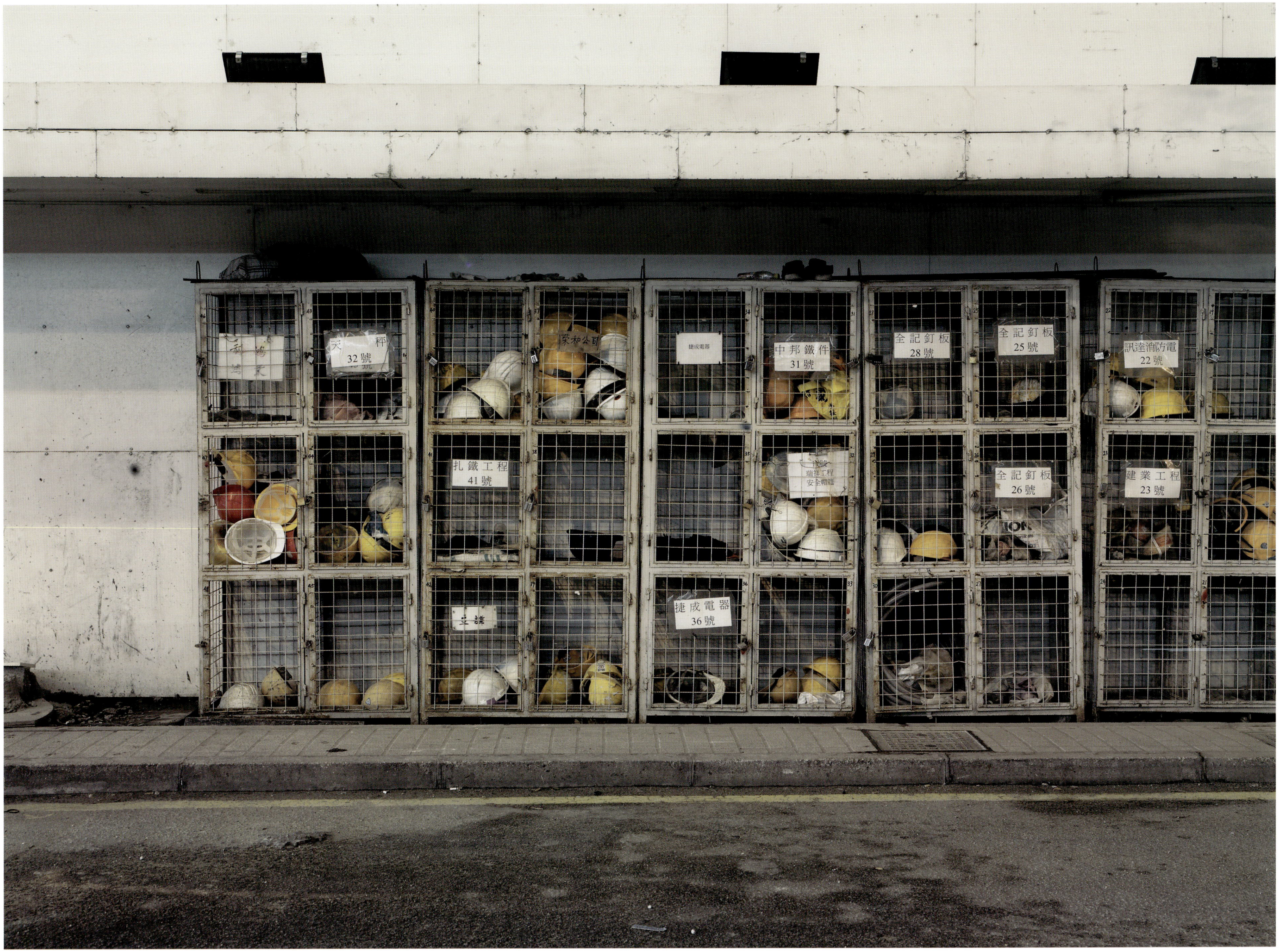
32號
扎鐵工程
41號
中邦鐵件
31號
全記釘板
28號
全記釘板
25號
訊達消防電
22號
全記釘板
26號
建業工程
23號
捷成電器
36號

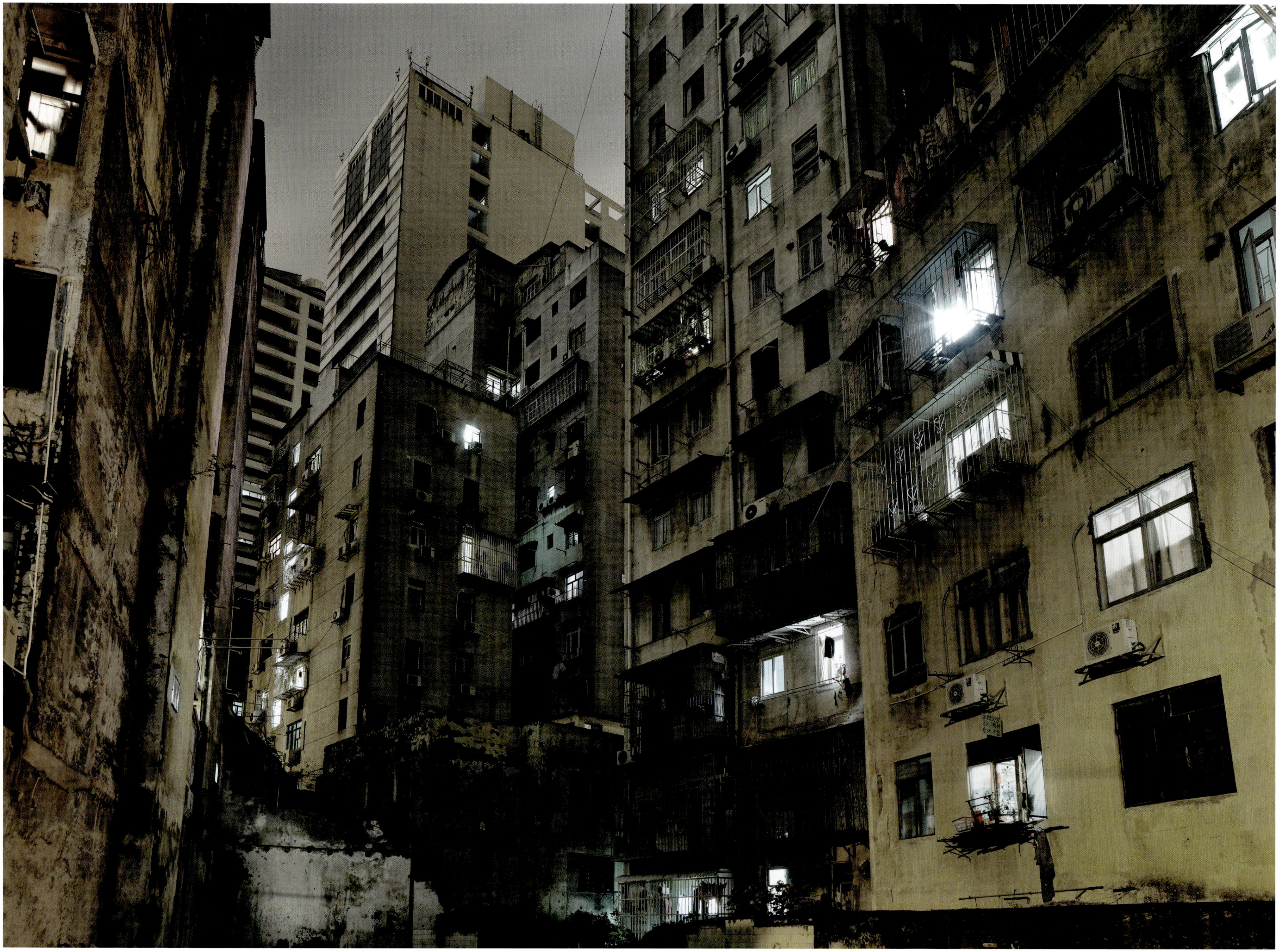

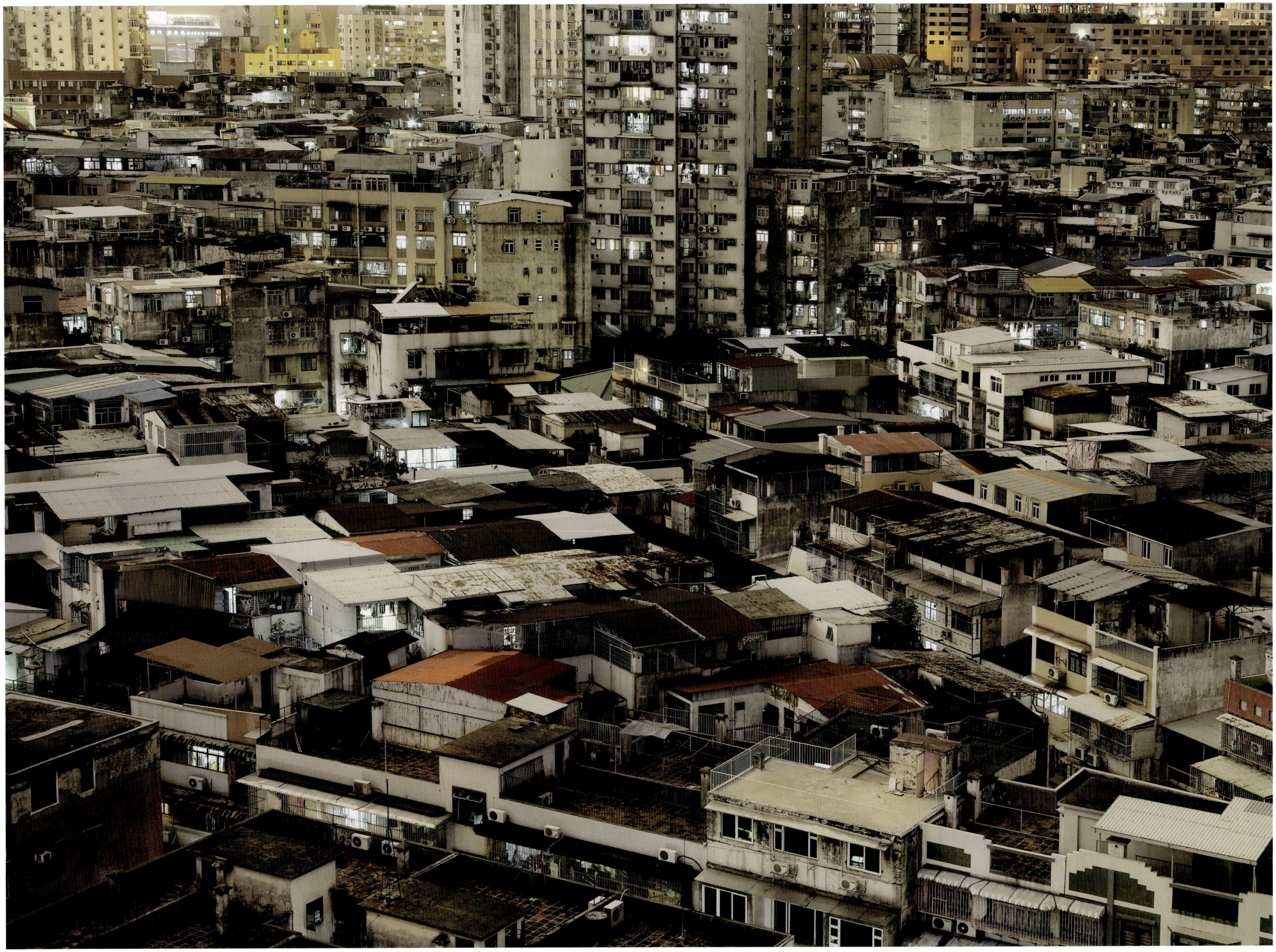

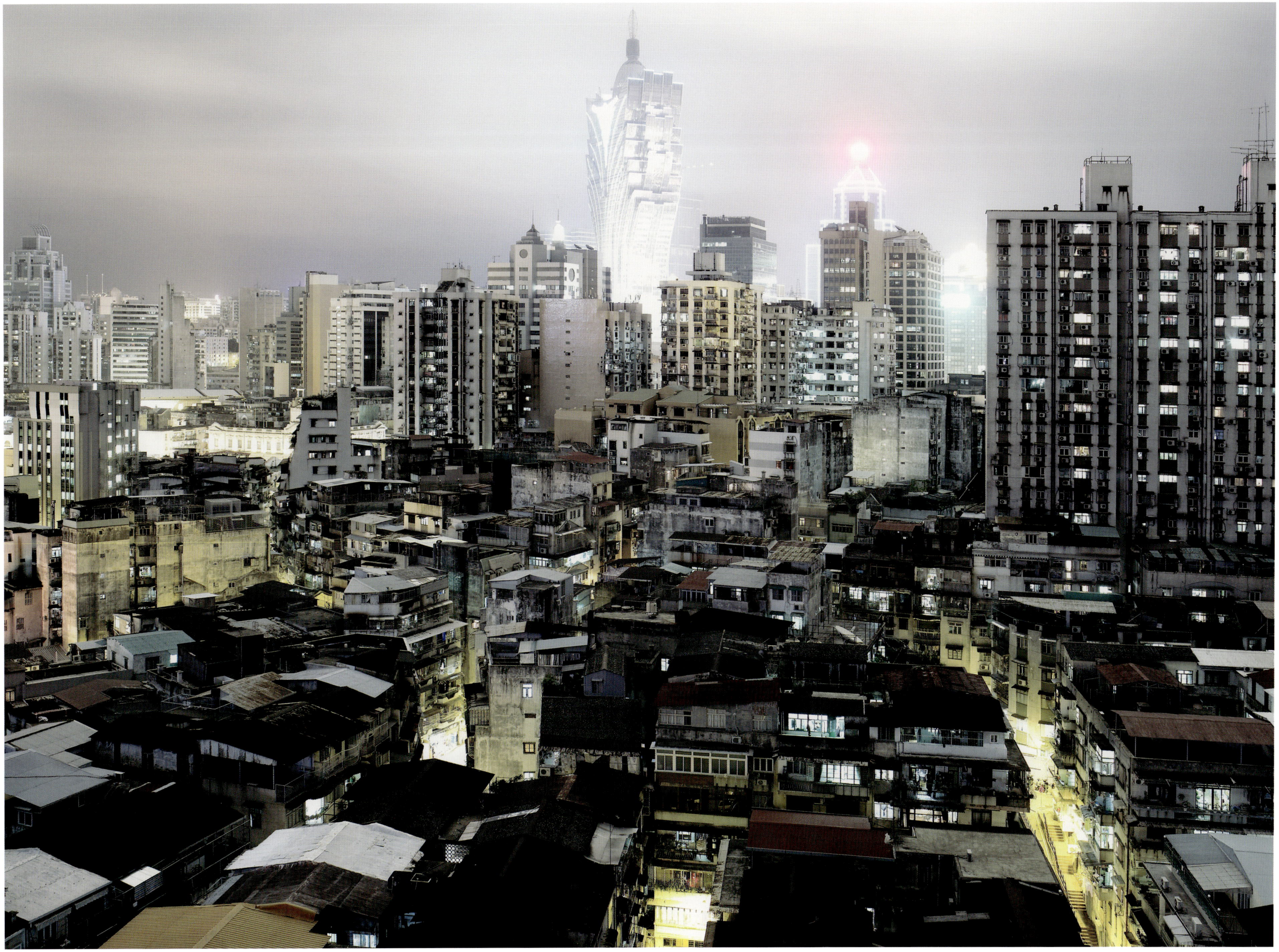

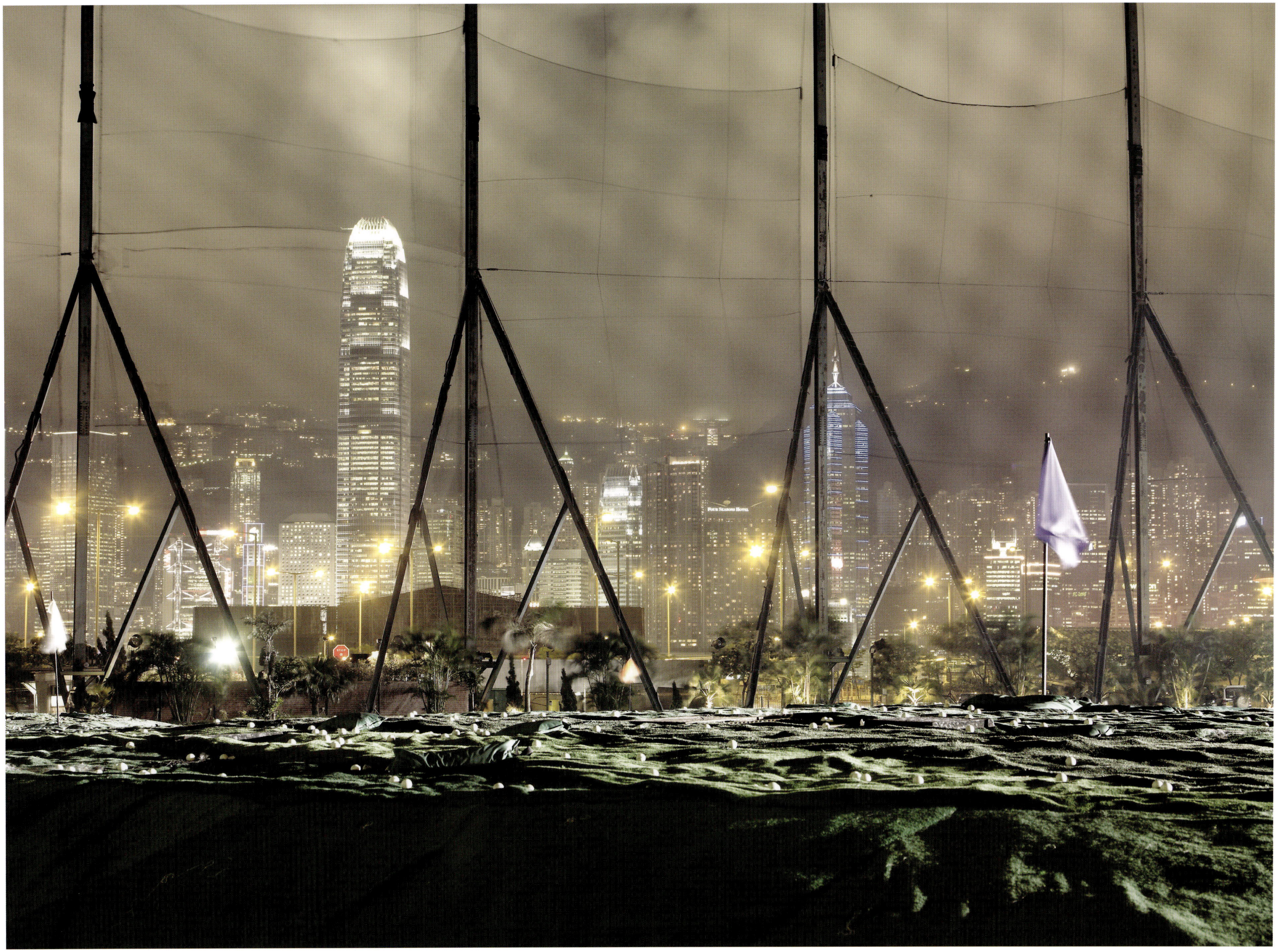

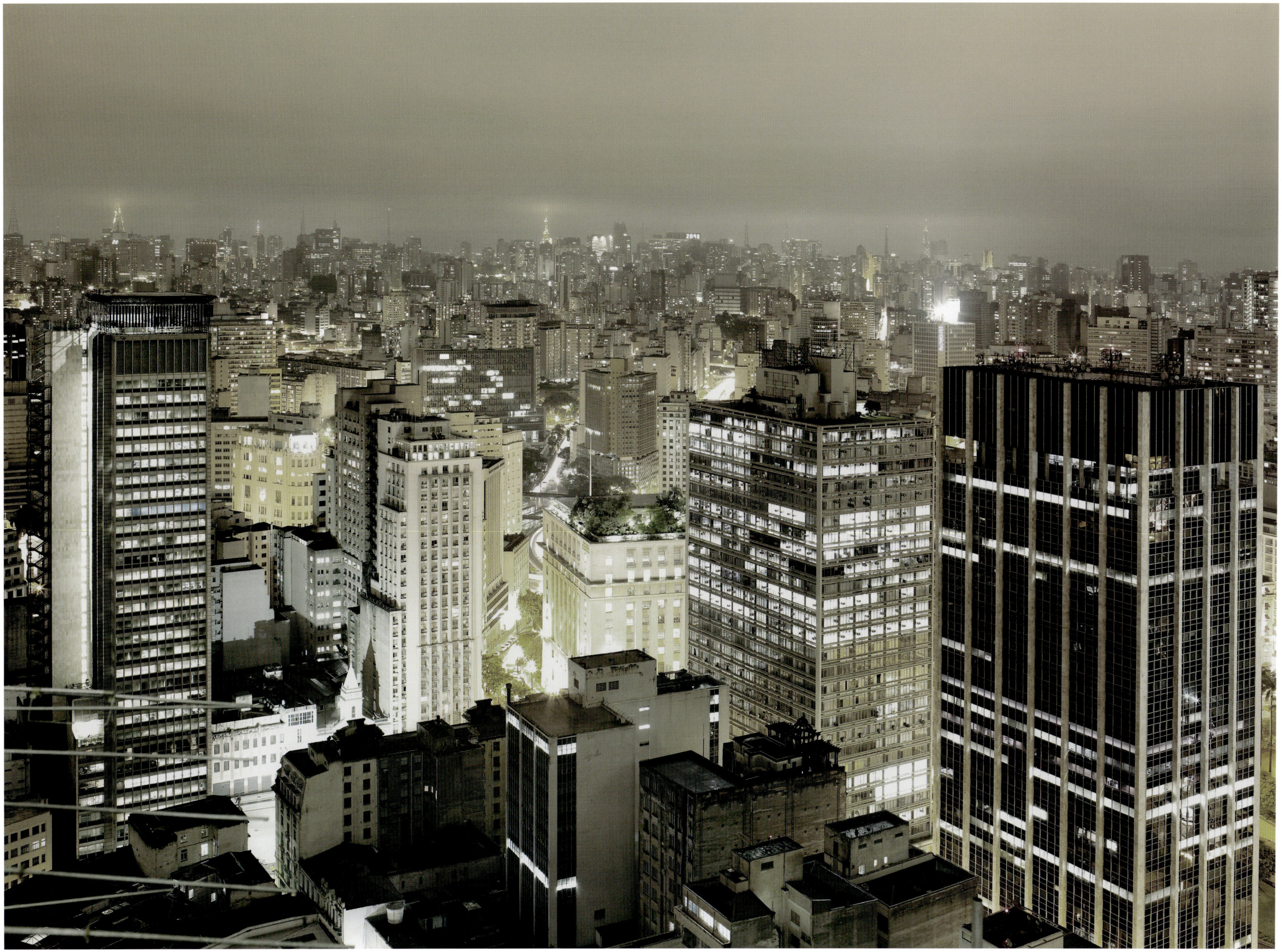

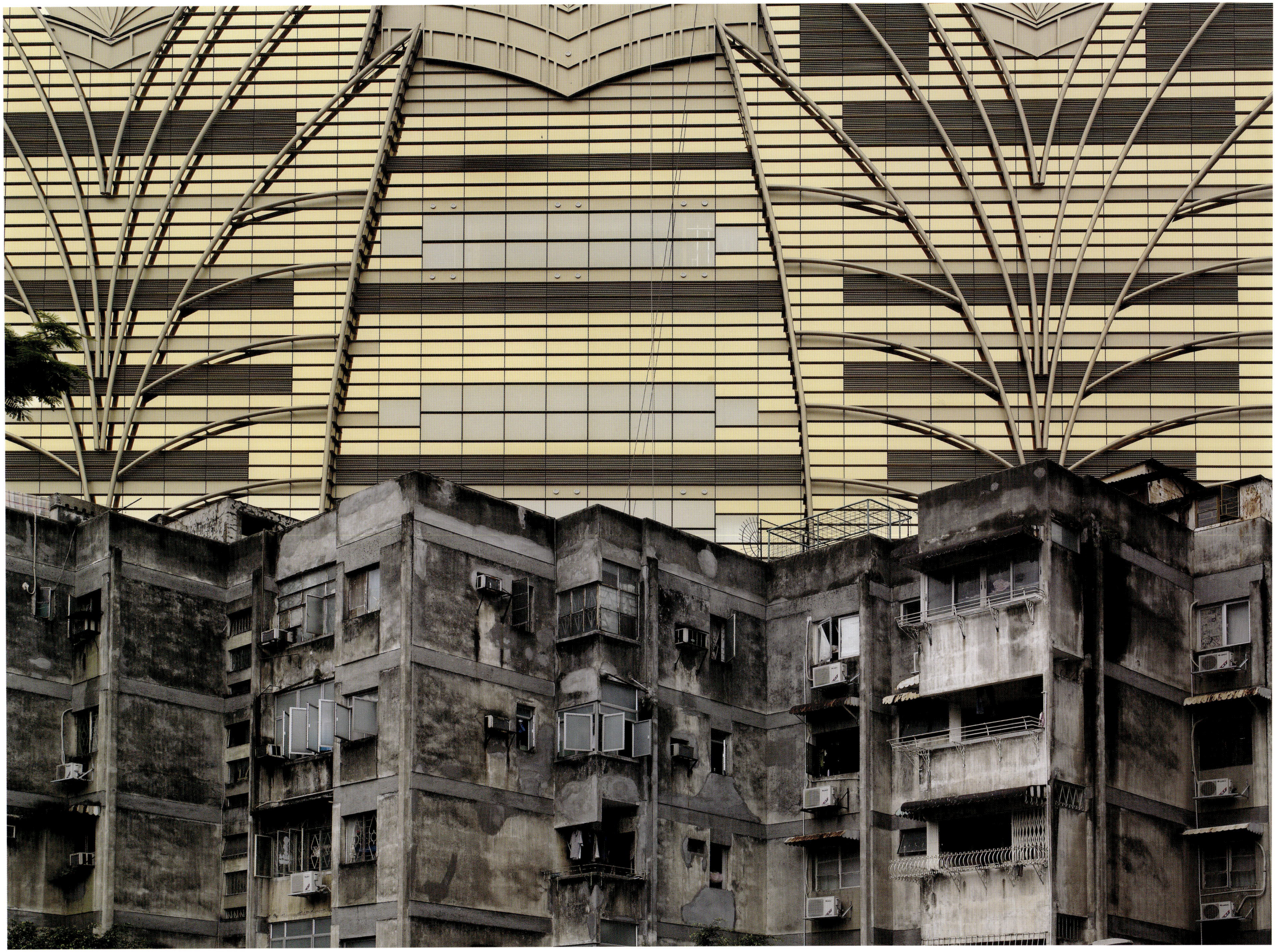

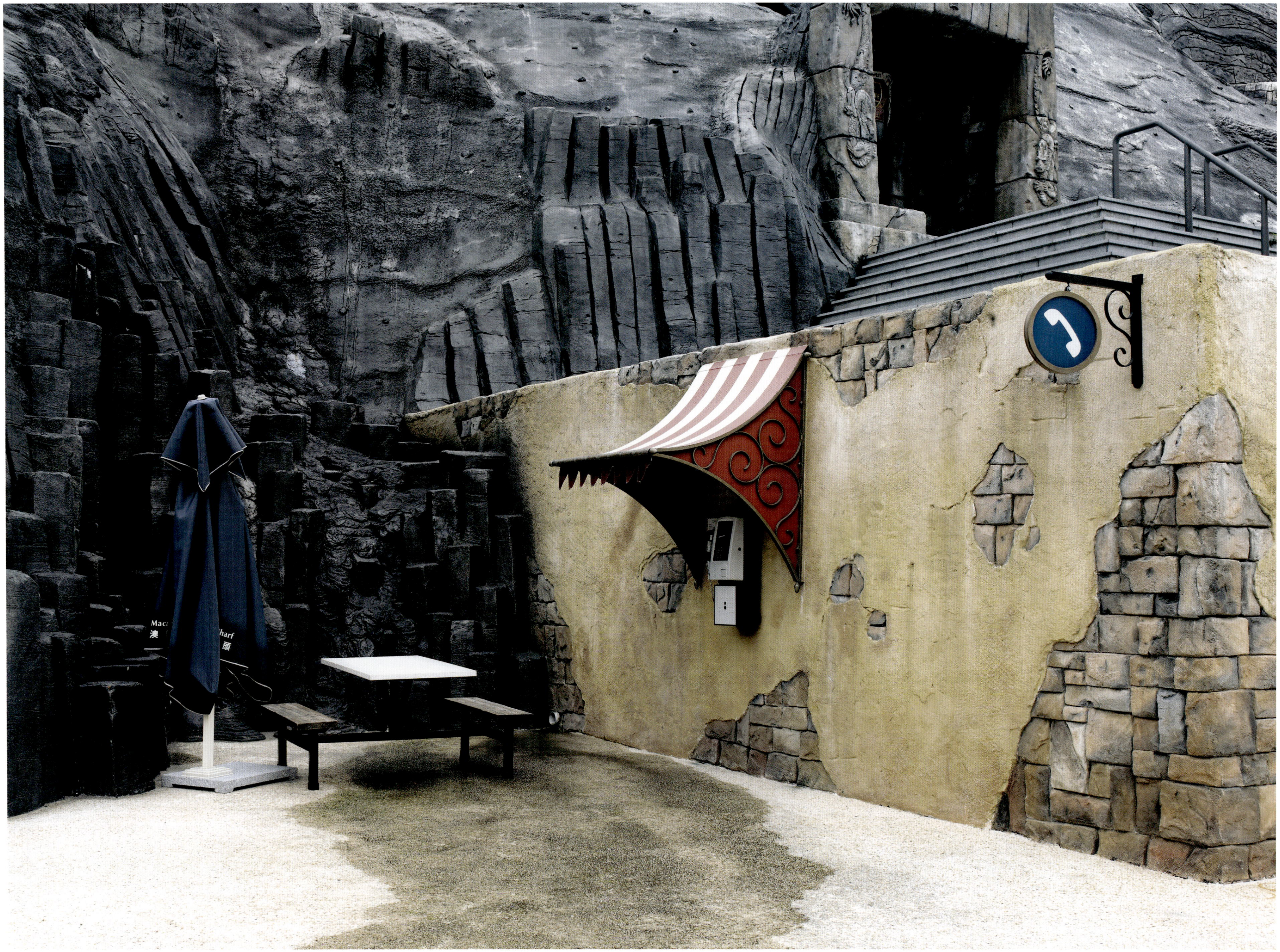

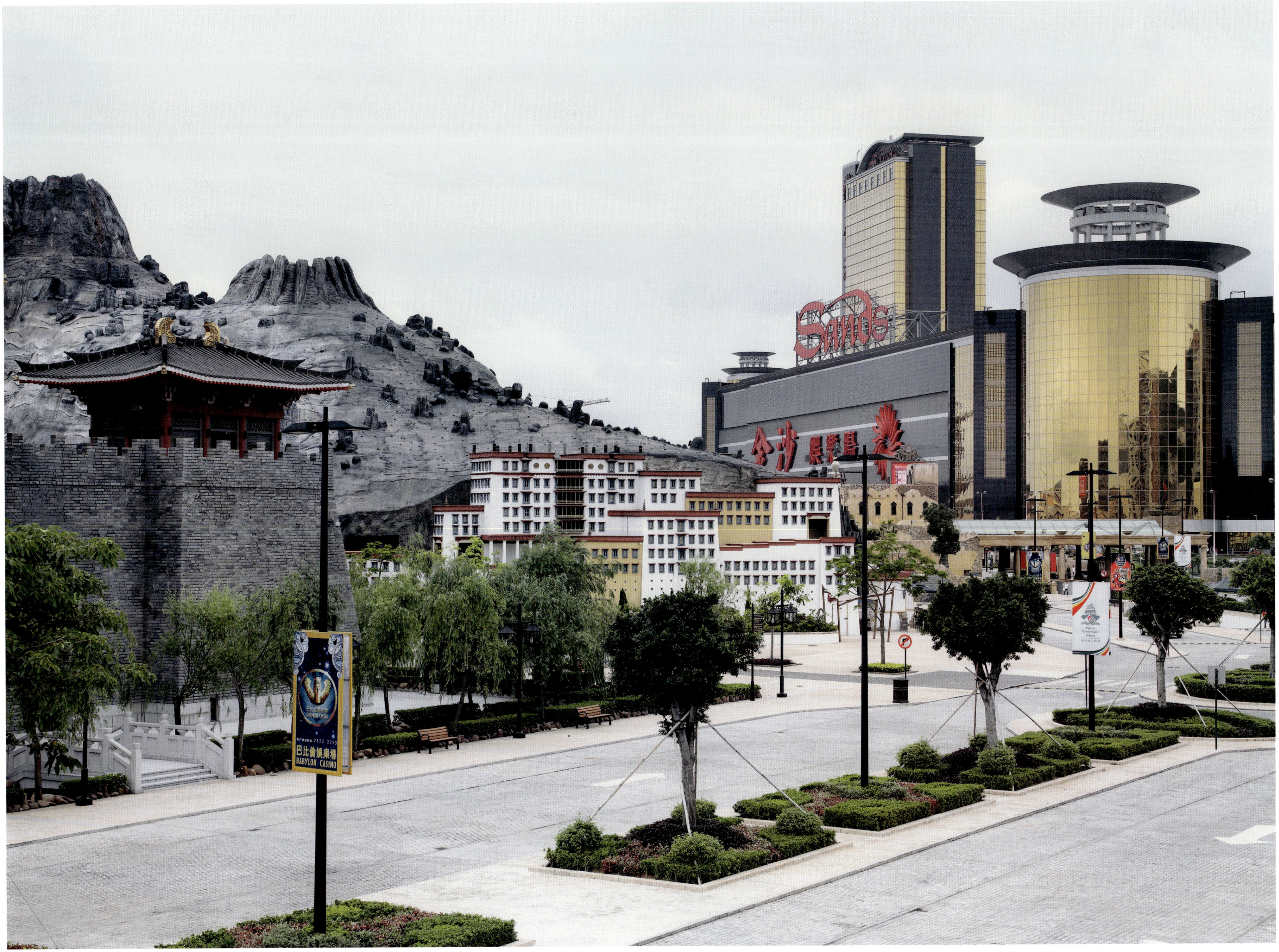
Sands
金沙
2823 2233
巴比倫娛樂場
BABYLON CASINO

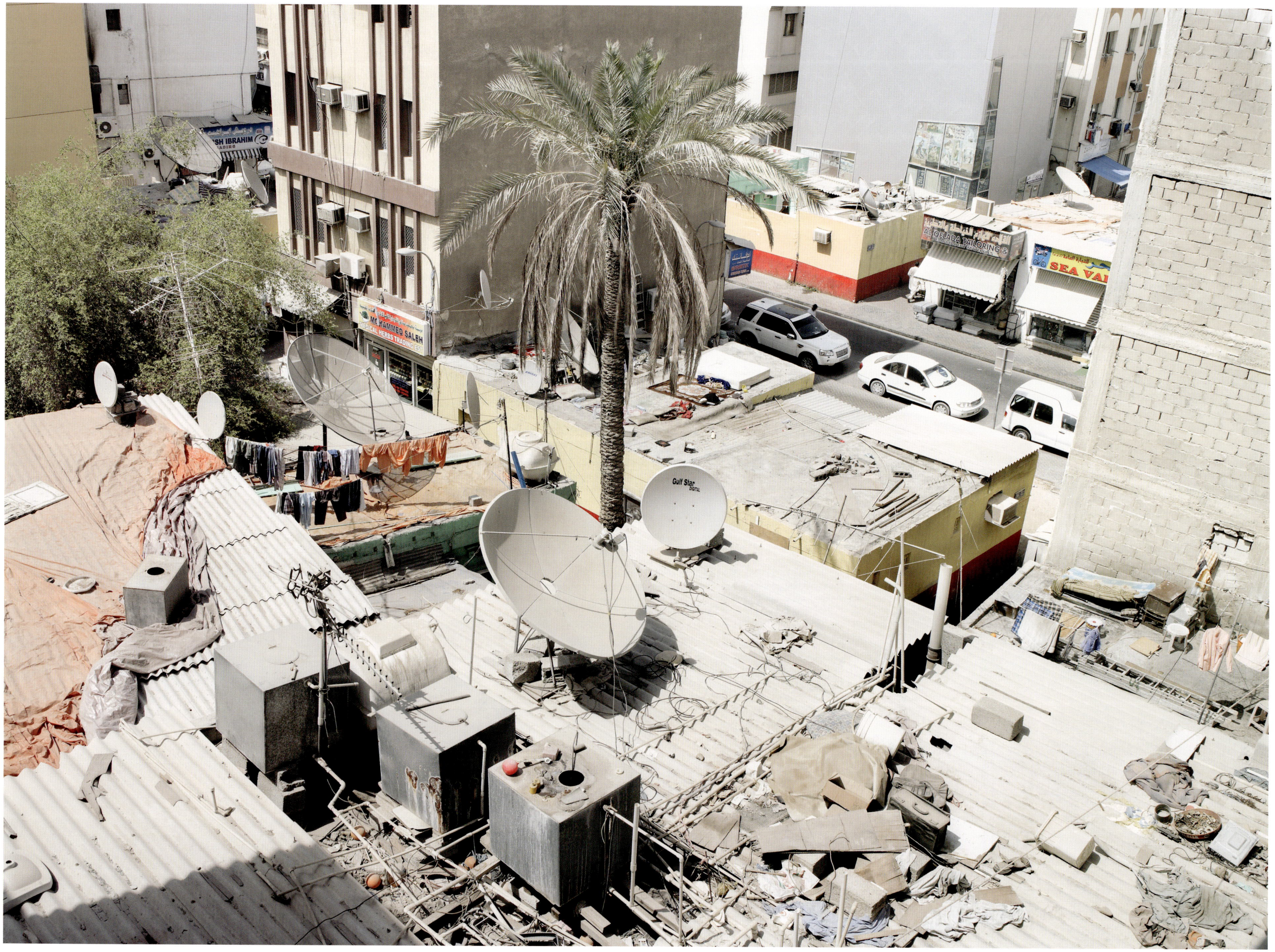
Gulf Star

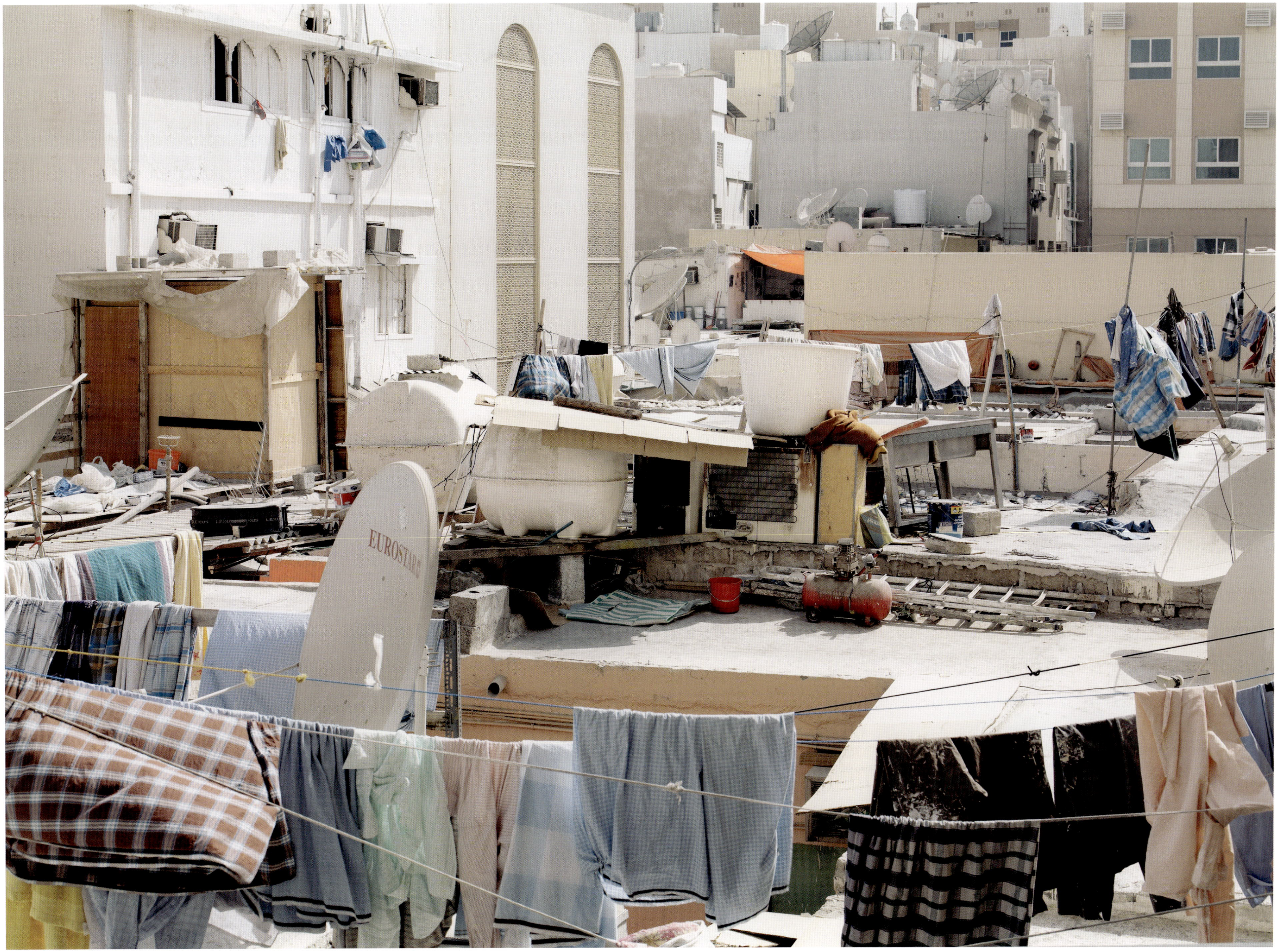
EUROSTAR

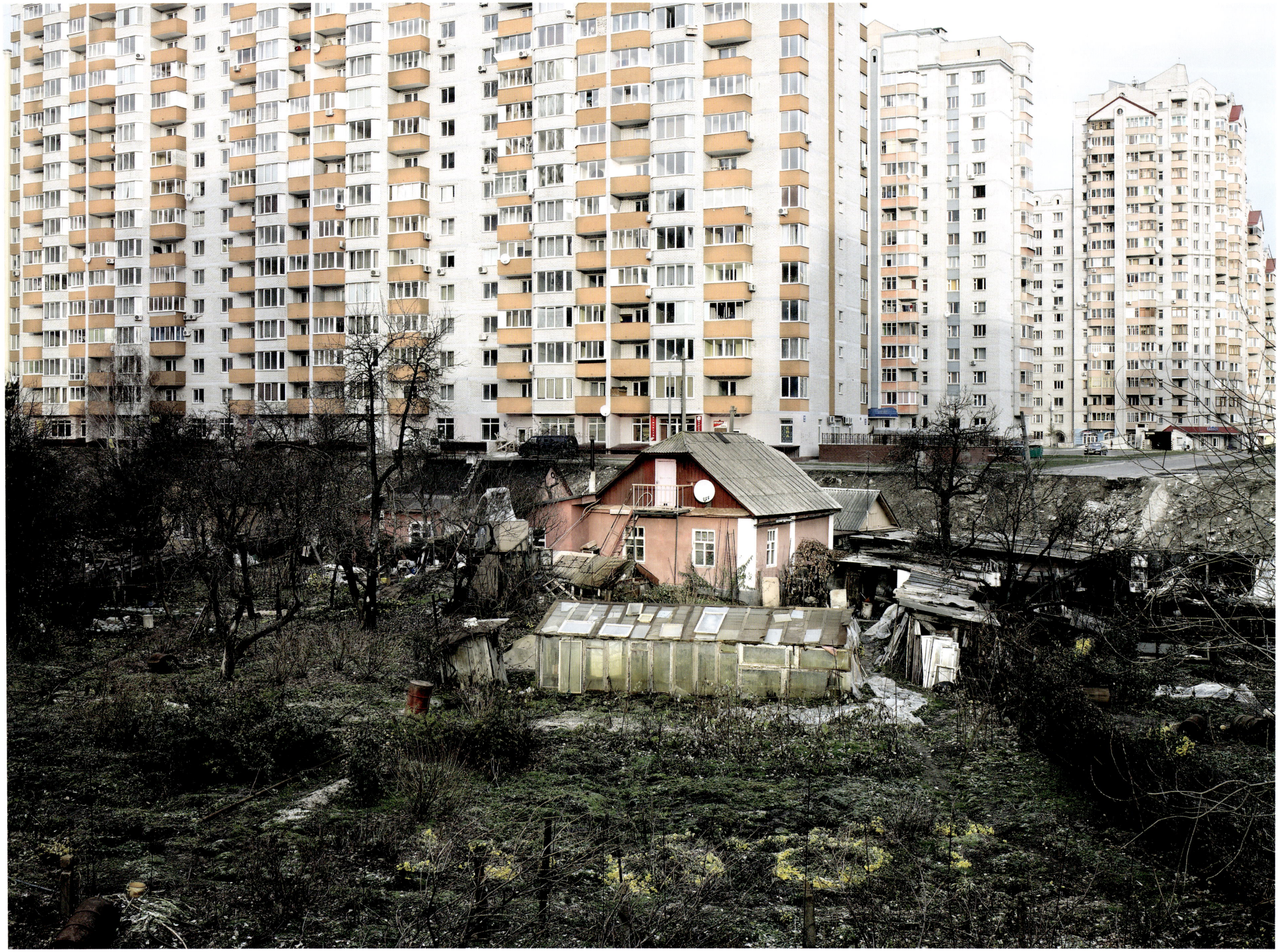

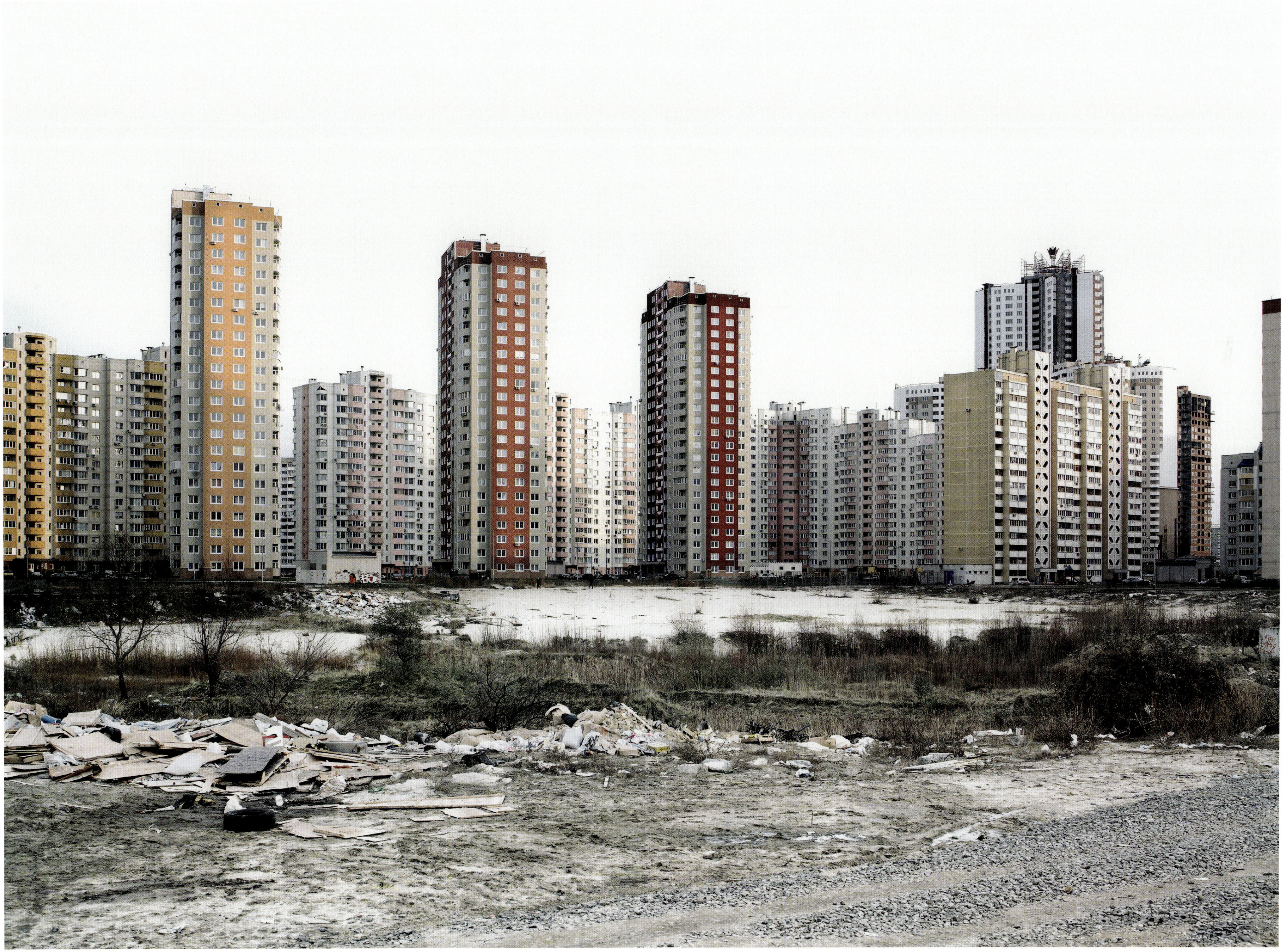

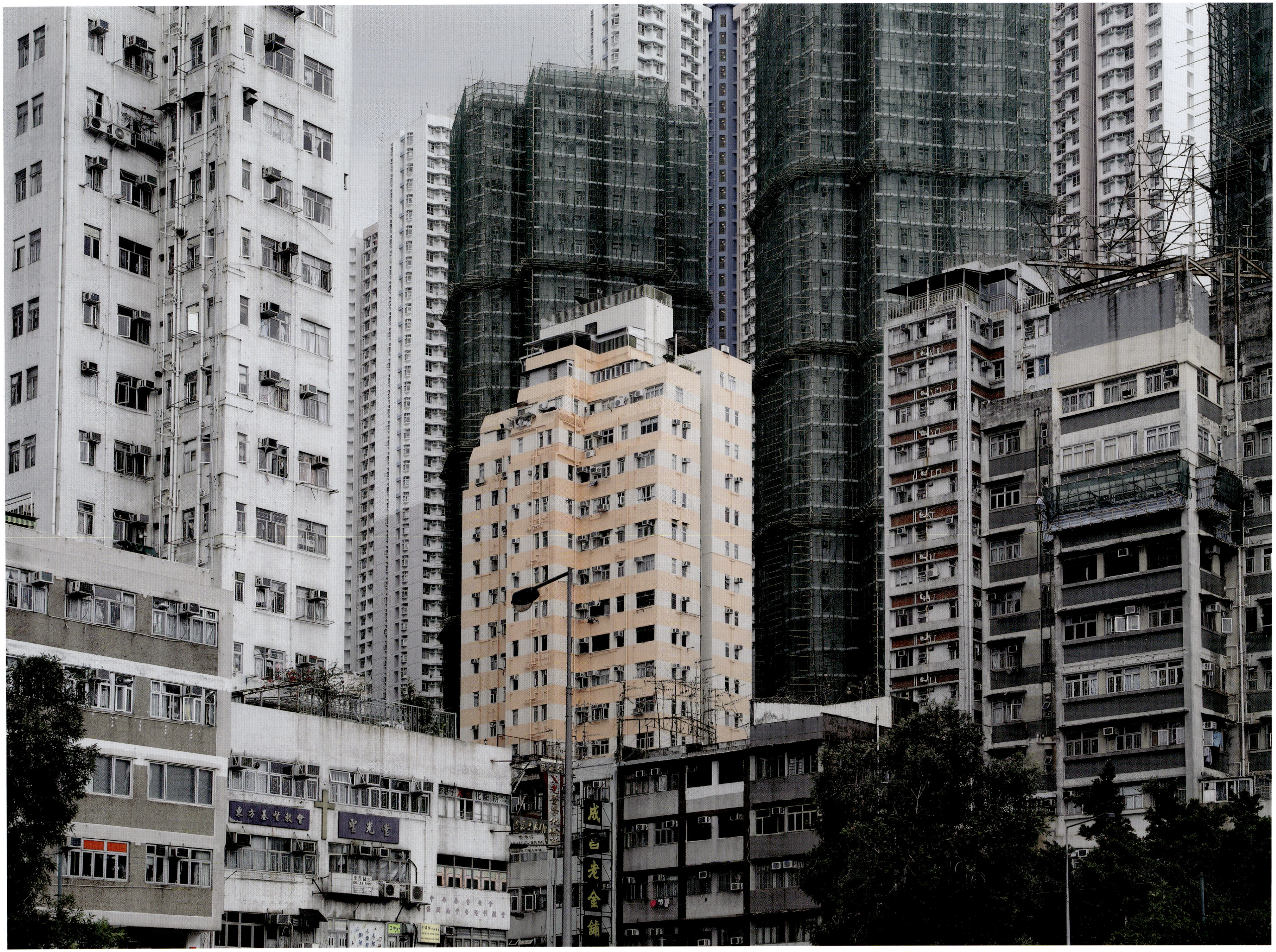
東方基督教會
聖光堂
李羨醫生
DR. LEE SHIN
中華基督教會
香港閩南堂香港仔副堂
成
白
光
金
舖

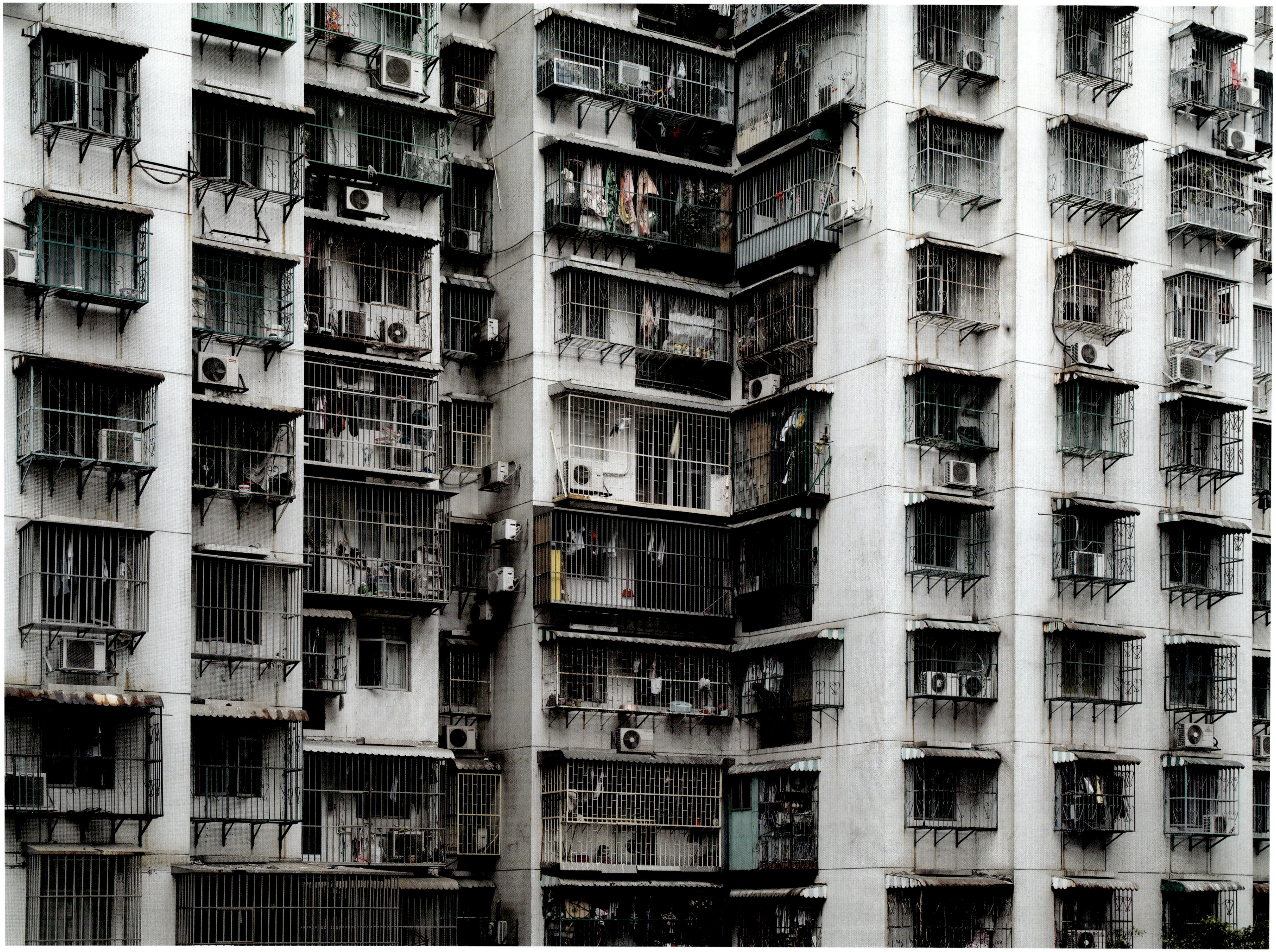

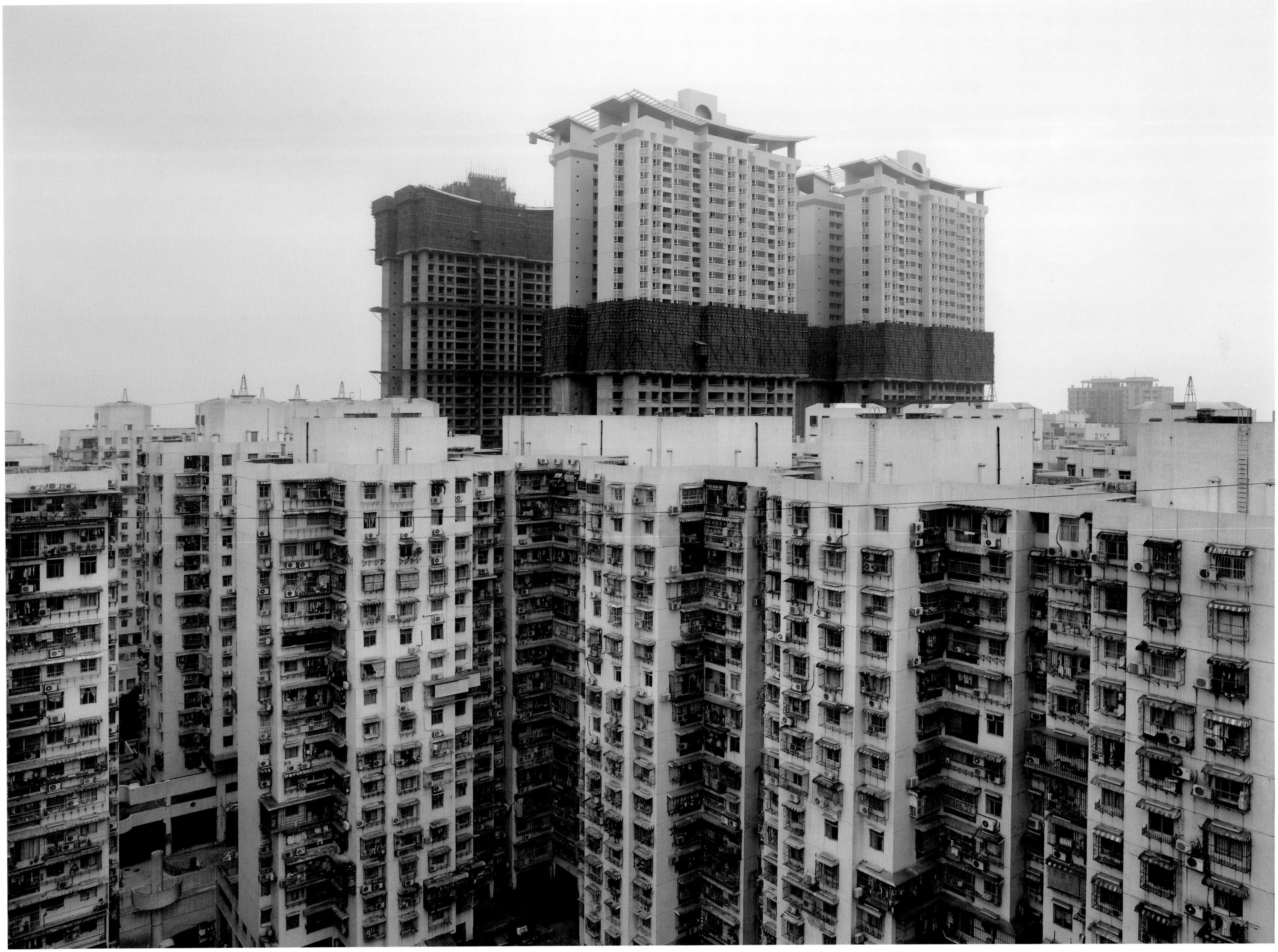

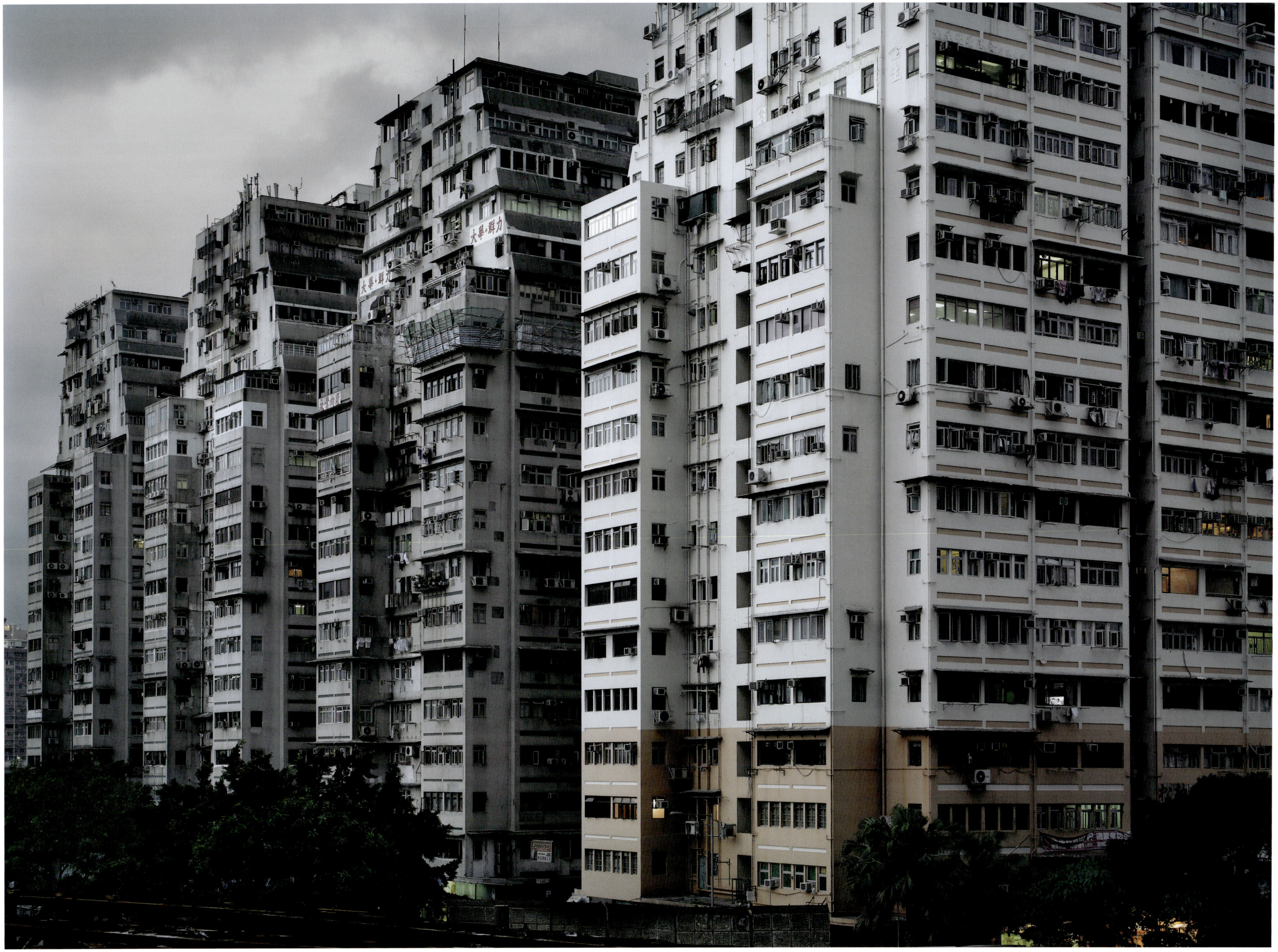
大華‧群力
大華‧群力

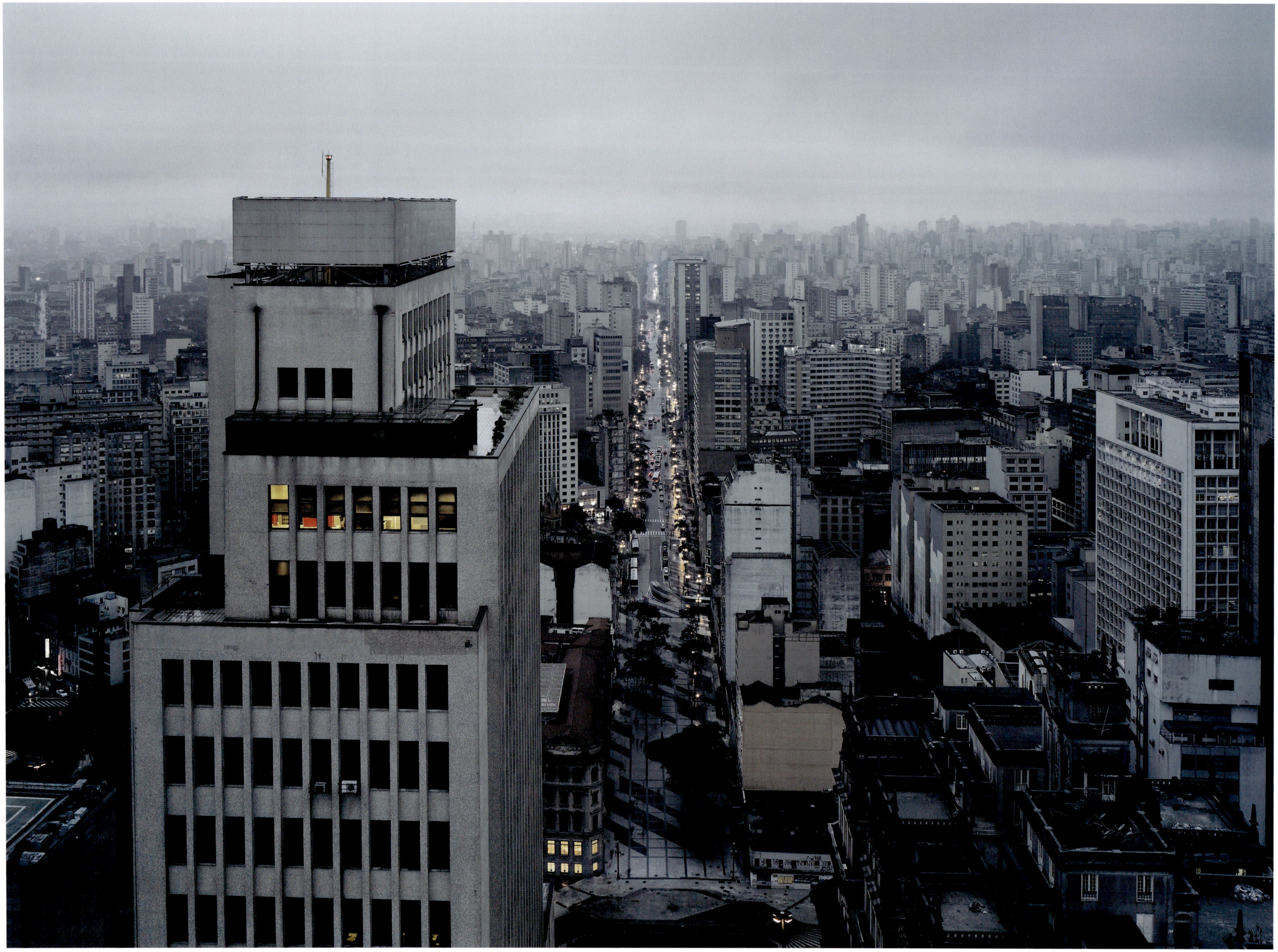

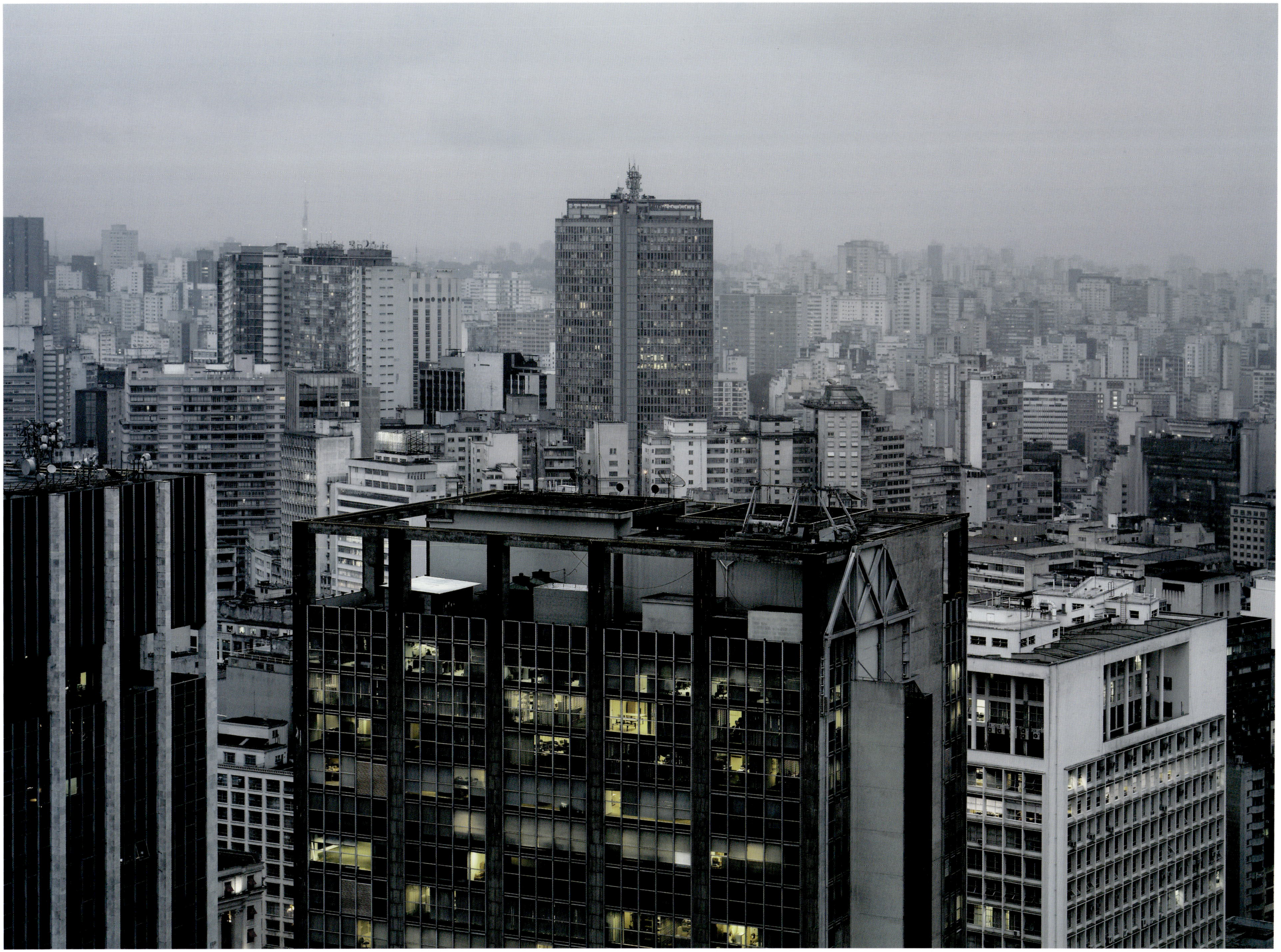

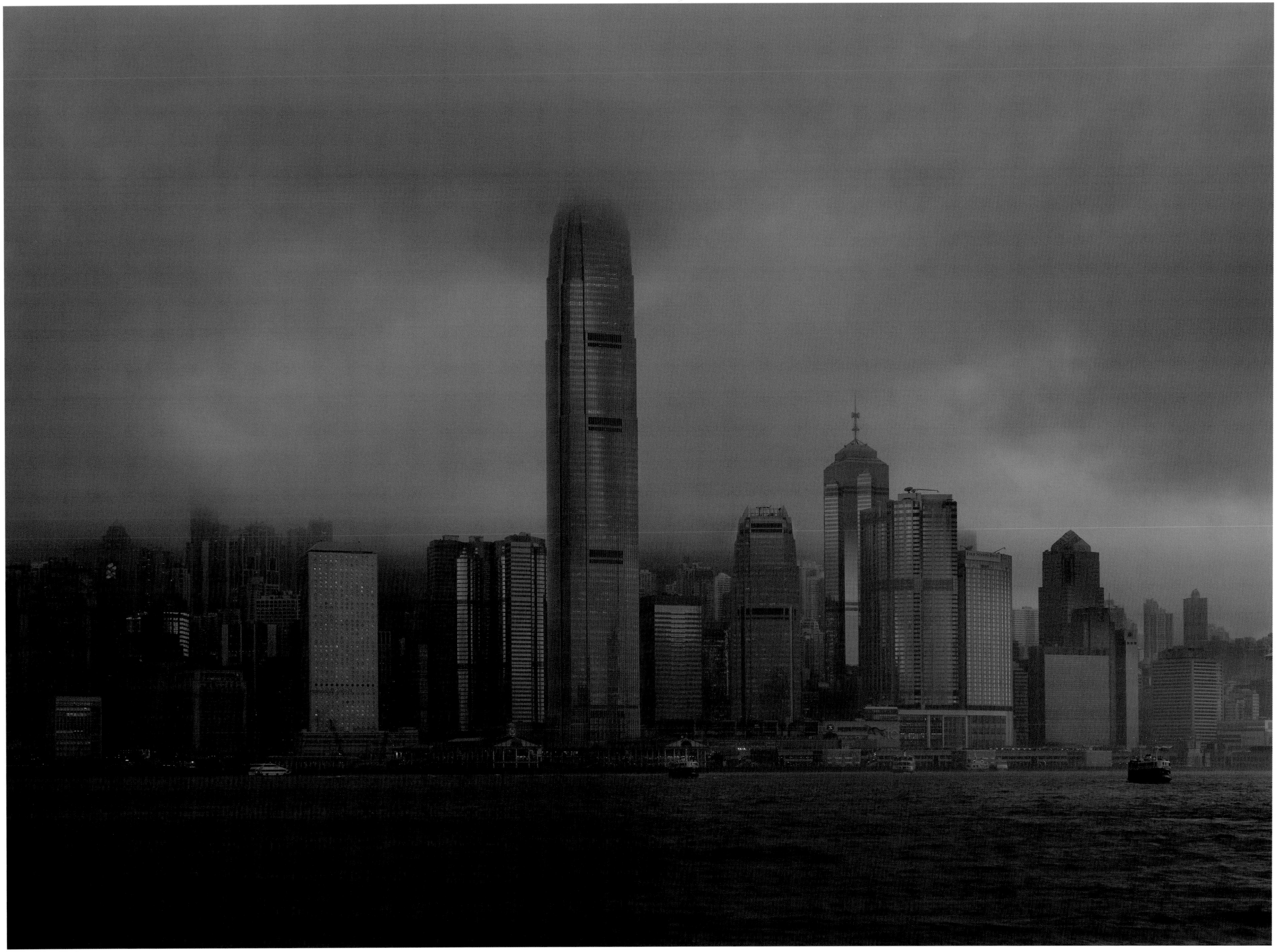

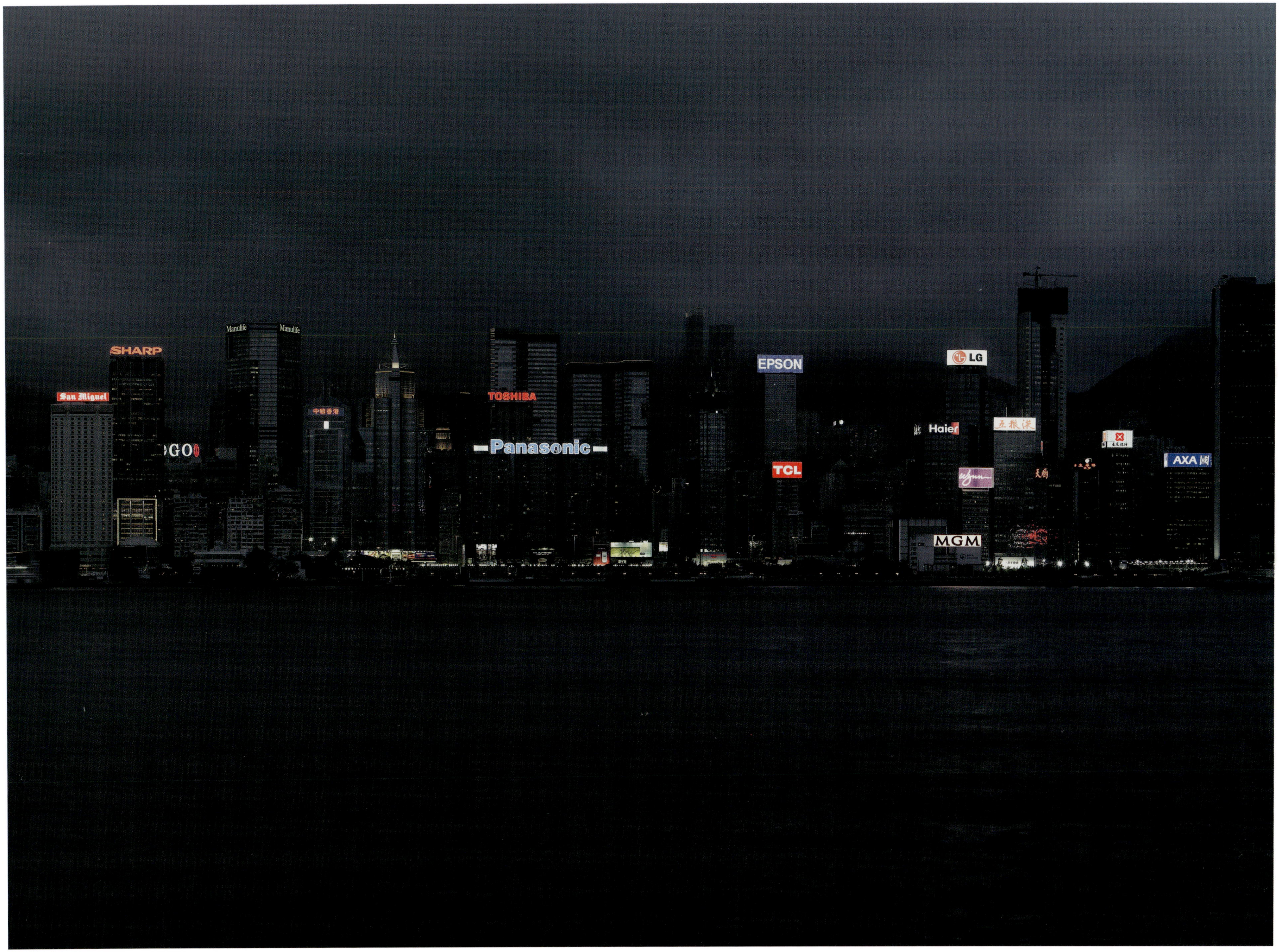
San Miguel
SHARP
Manulife
Manulife
TOSHIBA
Panasonic
EPSON
LG
Haier
TCL
MGM
AXA

# Plate List

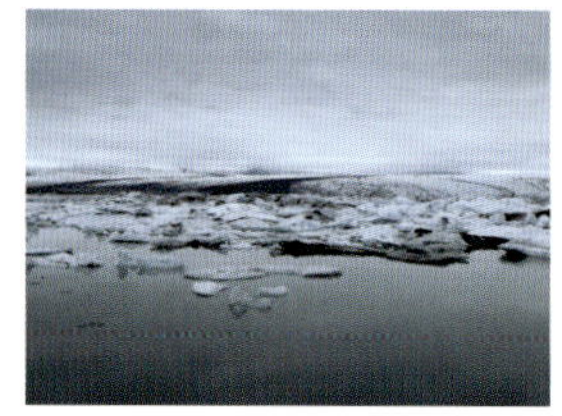
**11 /** Icebergs
*Iceland, 2007.*

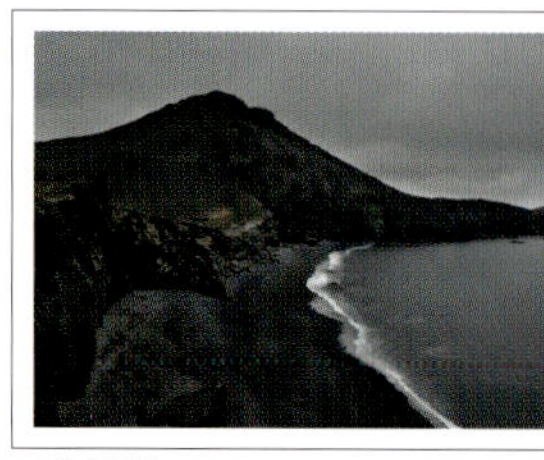
**12 /** Cliffs
*Iceland, 2007.*

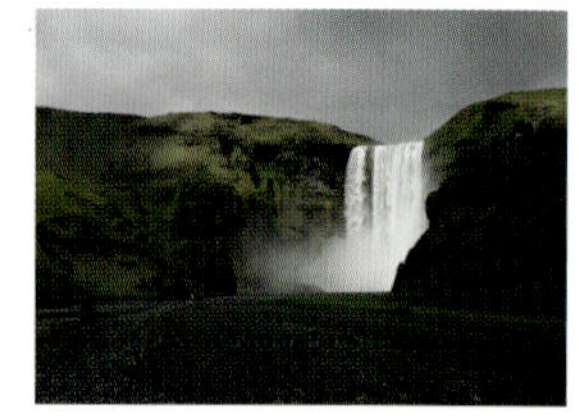
**13 /** Waterfall
*Iceland, 2007.*

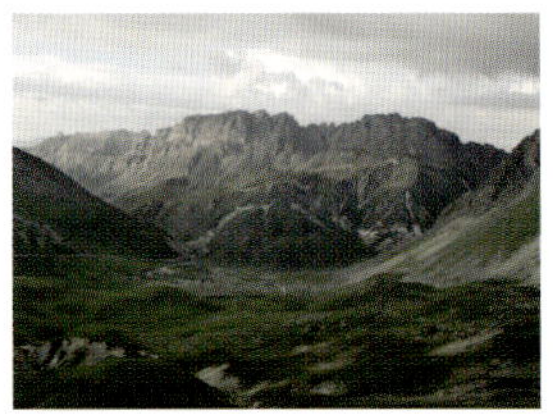
**15 /** Mountains
*France, 2006.*

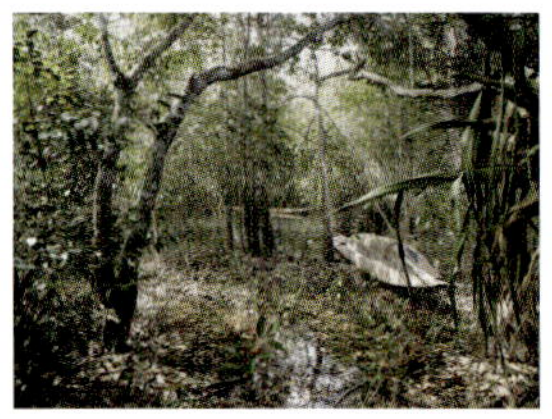
**16 /** Pirogues
*Amazonian forest, Brazil, 2006.*

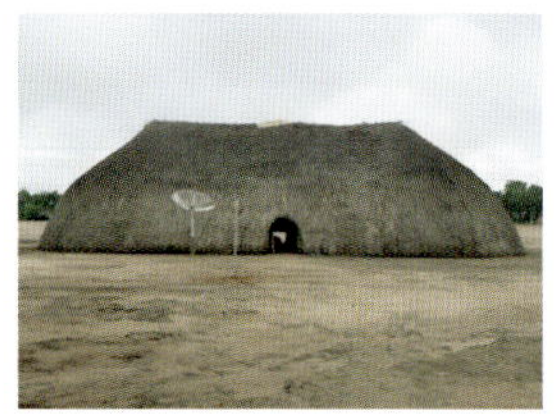
**17 /** Xingu hut
*Amazonian forest, Brazil, 2006.*

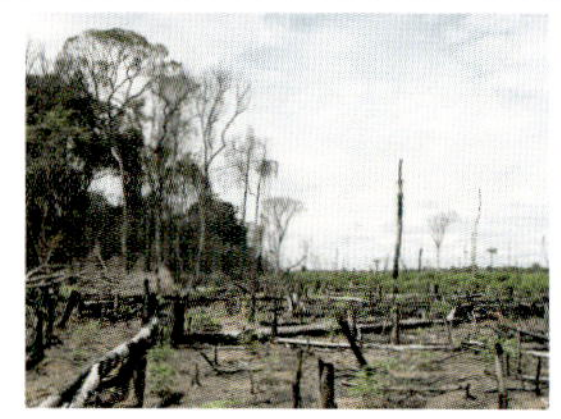
**19/** Deforestation
*Brazil, 2006.*

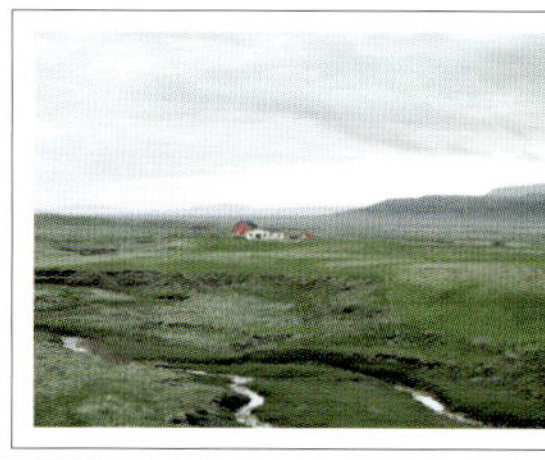
**20 /** Prairie house
*Iceland, 2007.*

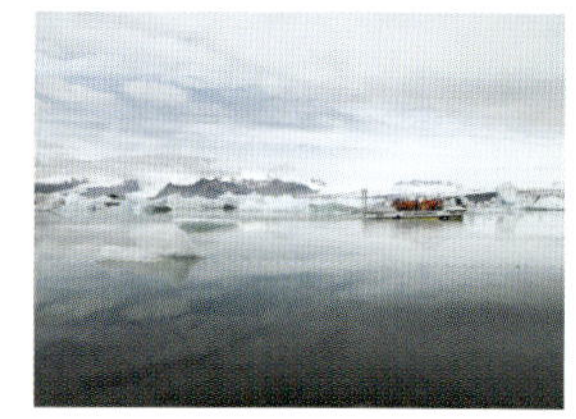
**21 /** Tourists
*Iceland, 2007.*

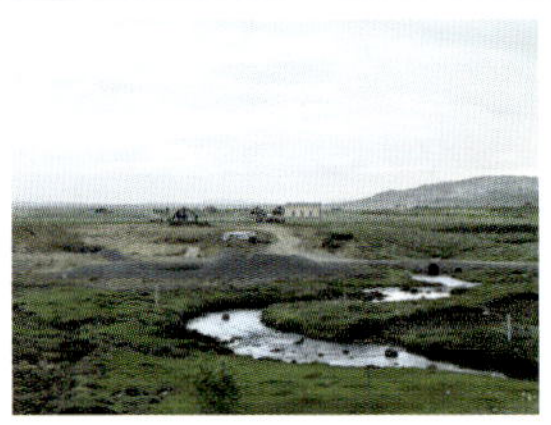
**23 /** Farm
*Iceland, 2007.*

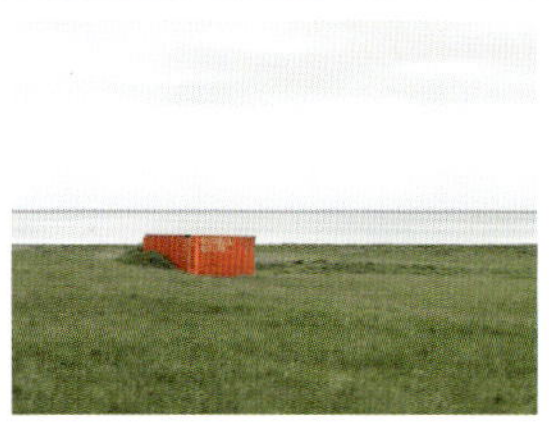
**24 /** Container
*Iceland, 2007.*

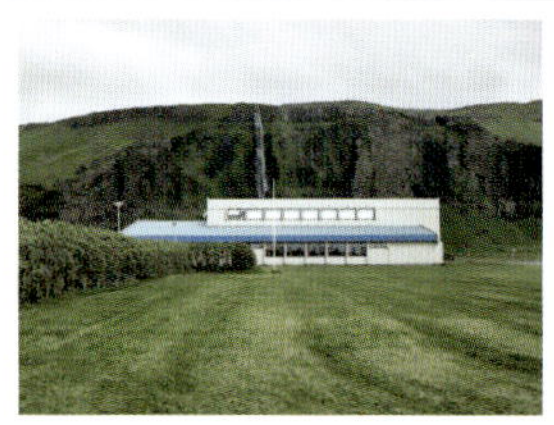
**25 /** Home for youth
*Iceland, 2007.*

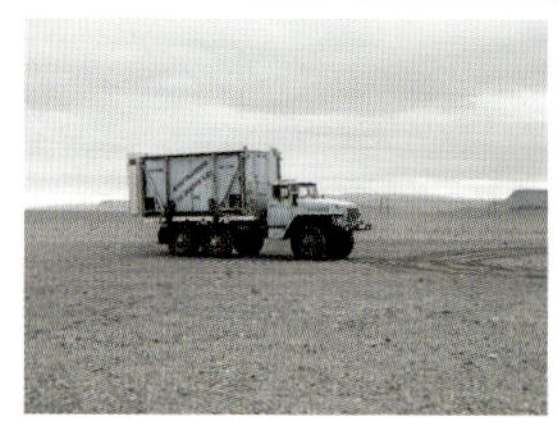
**27 /** Broken-down truck
*Iceland, 2007.*

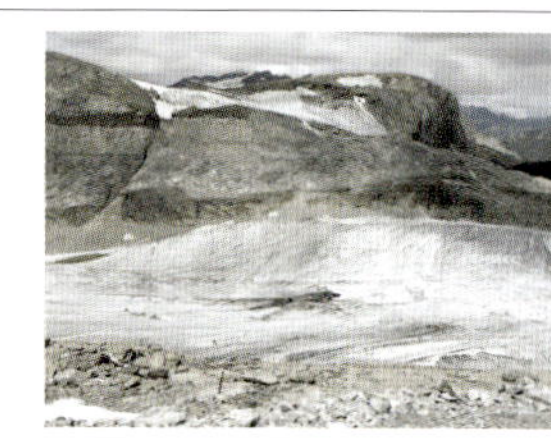
**29 /** Ski lift
*France, 2006.*

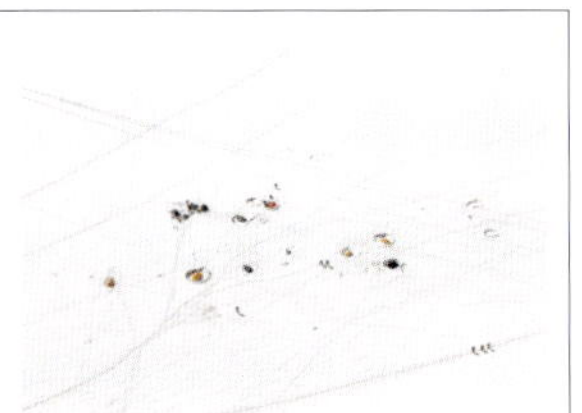
**31 /** Base camp on Mont Blanc
*France, 2006.*

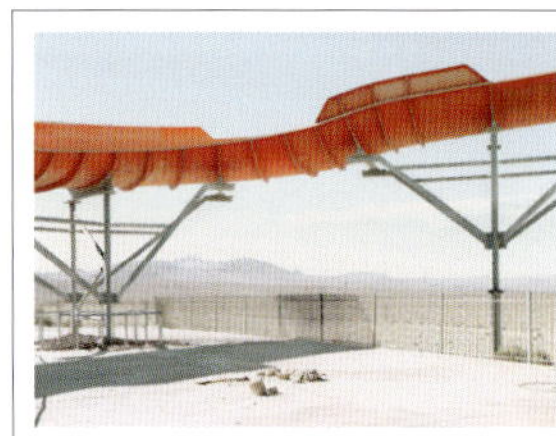
**32 /** Aquatic amusement park
*USA, 2007.*

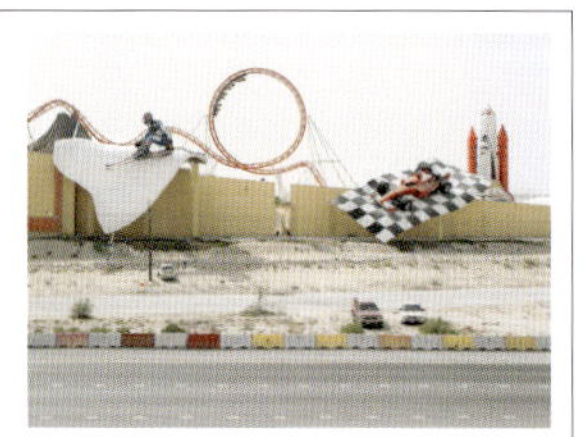
**33 /** Billboards for Dubailand
*United Arab Emirates, 2008.*

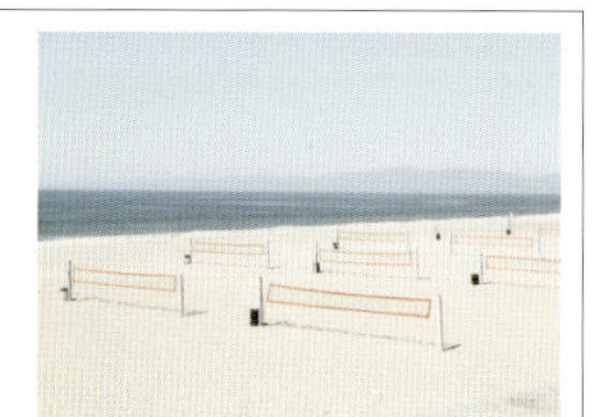
**35 /** Volleyball court
*Venice Beach, California, USA, 2008.*

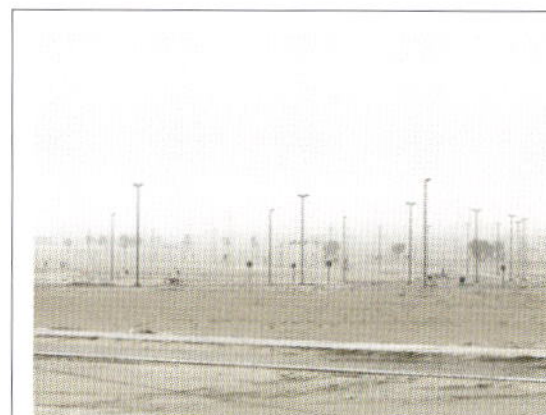
**36 /** Road construction in the desert
*United Arab Emirates, 2008.*

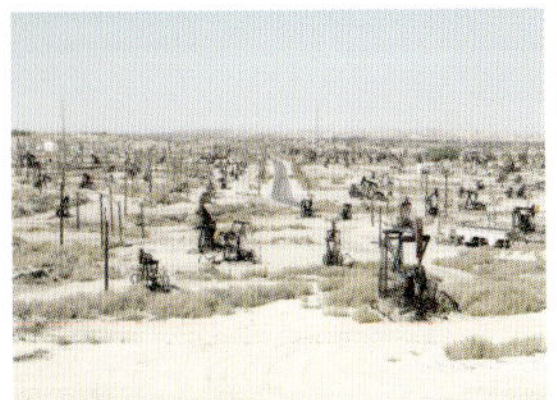
**39 /** Petroleum field 1
USA 2007

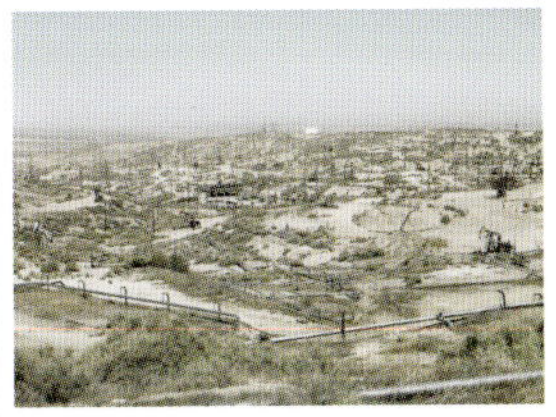
**40 /** Petroleum field 2
USA 2007

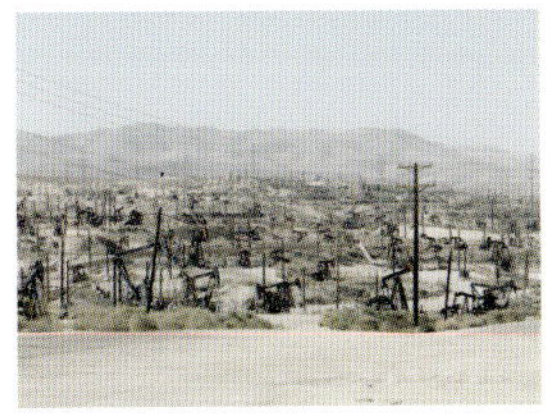
**41 /** Petroleum field 3
USA 2007

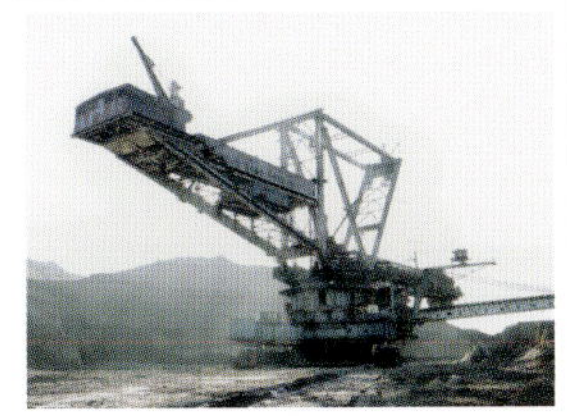
**43 /** Mining
Ukraine 2007

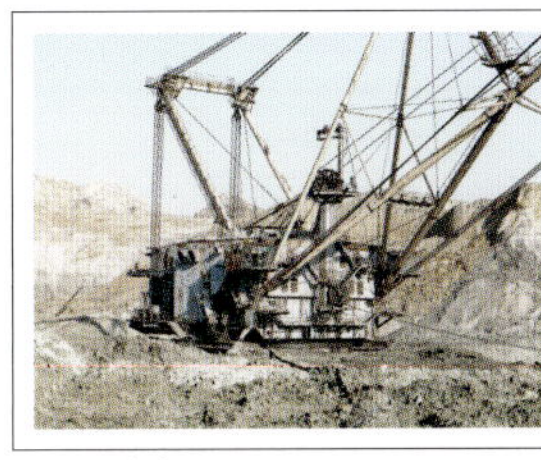
**44 /** Mining
Ukraine 2007

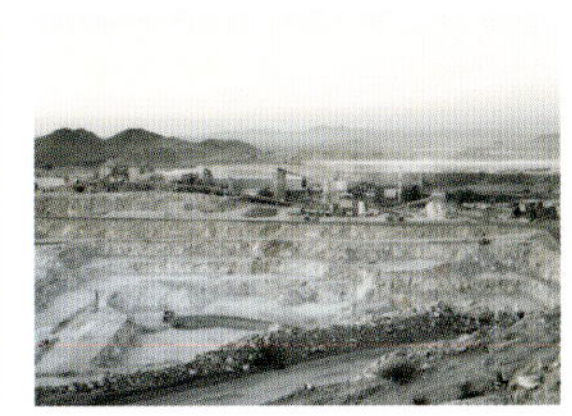
**45 /** Cement works quarry
USA 2007

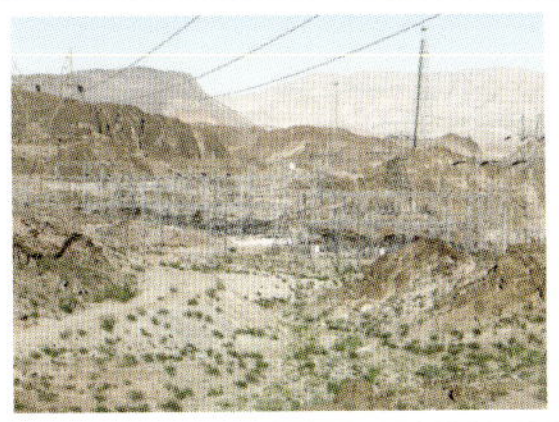
**47 /** Power substation
USA 2007

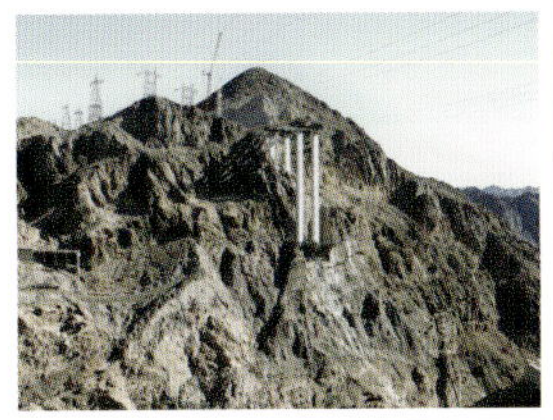
**49 /** Bridge under construction
USA 2007

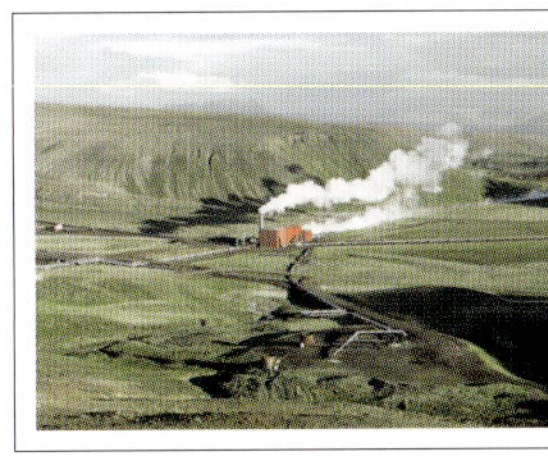
**50 /** Geothermal plant 1
Iceland 2007

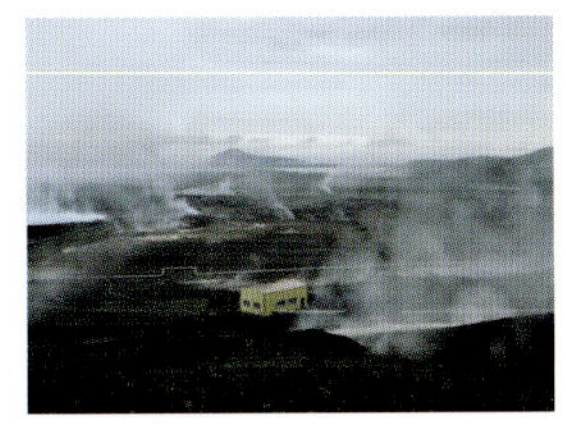
**51 /** Geothermal plant 2
Iceland 2007

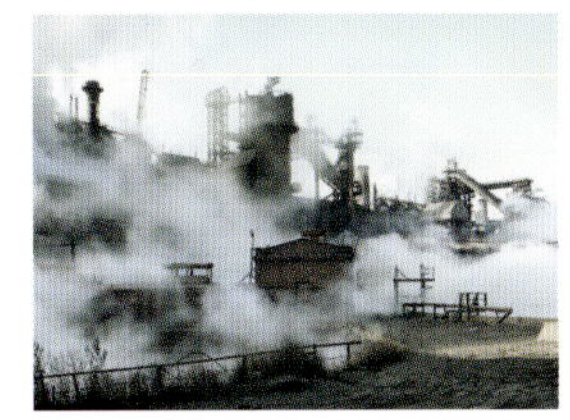
**53 /** Steel mill 1
Ukraine 2007

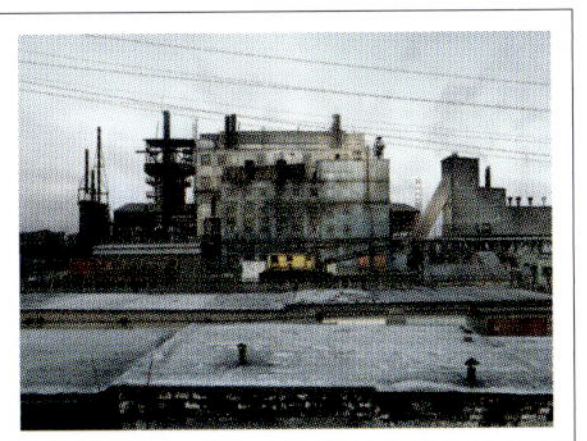
**55 /** Abandoned factory
Ukraine 2007

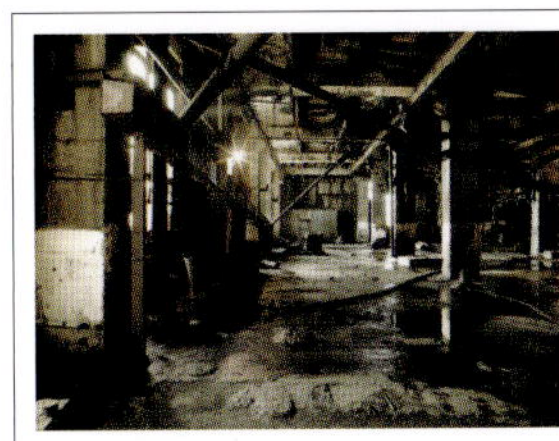
**56 /** Manganese mine 1
Ukraine 2007

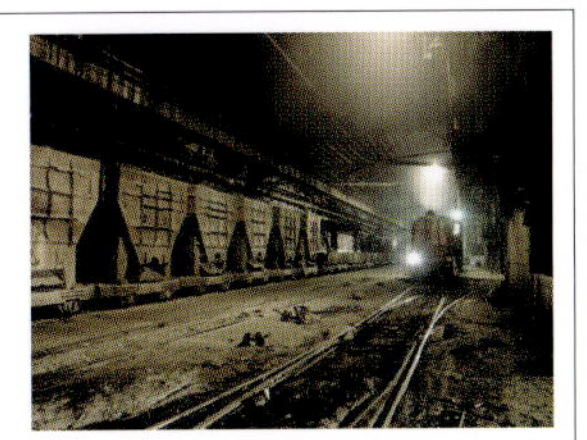
**57 /** Steel mill 2
Ukraine 2007

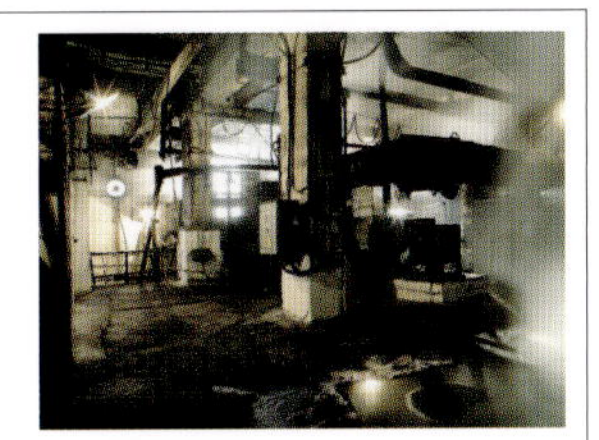
**59 /** Manganese mine 2
Ukraine 2007

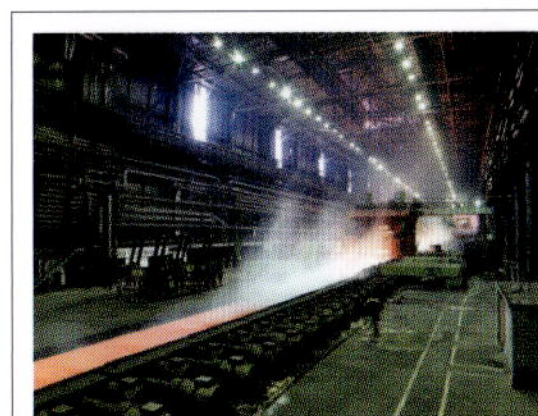
**60 /** Steel mill 3
Ukraine 2007

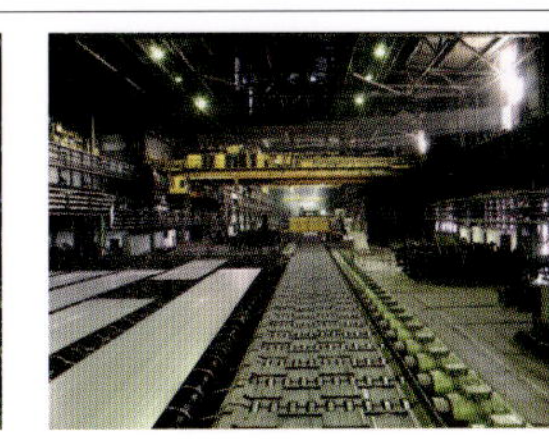
**61 /** Steel mill 4
Ukraine 2007

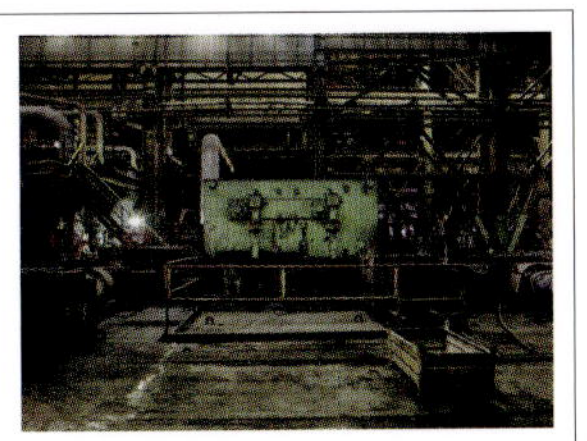
**63 /** Steel mill 5
Ukraine 2007

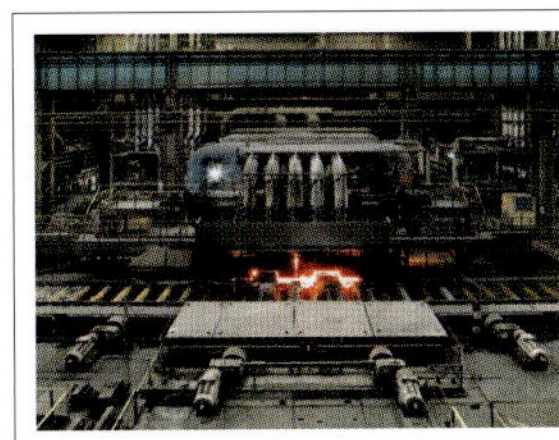
**64 /** Steel mill 6
Ukraine 2007

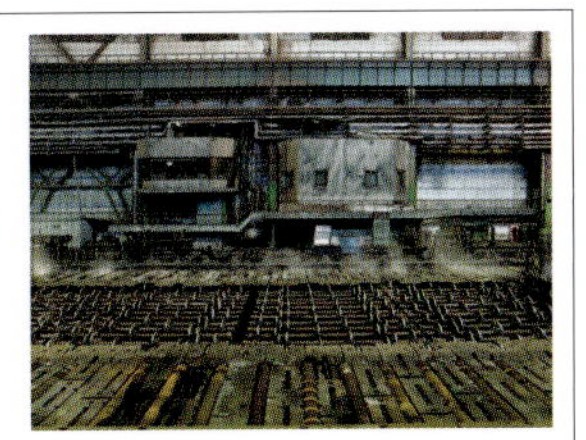
**65 /** Steel mill 7
Ukraine 2007

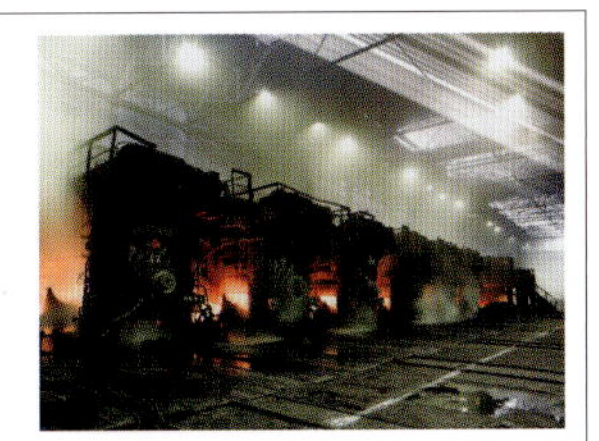
**67 /** Steel mill 8
Ukraine 2007

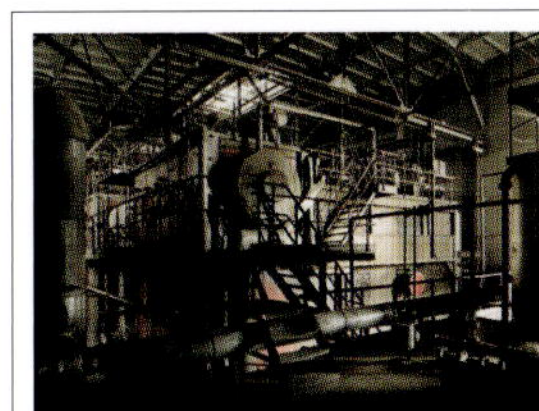
**68 /** Incineration plant 1
Ukraine 2007

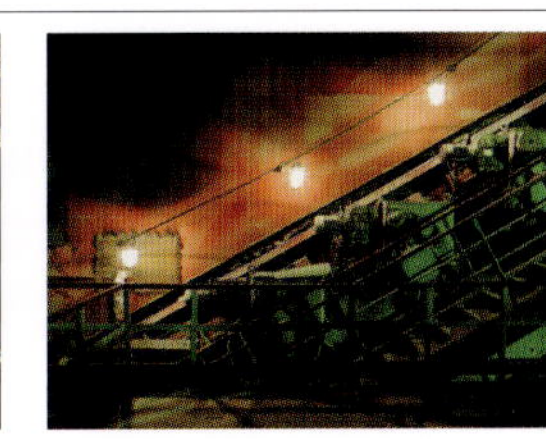
**69 /** Incineration plant 2
Ukraine 2007

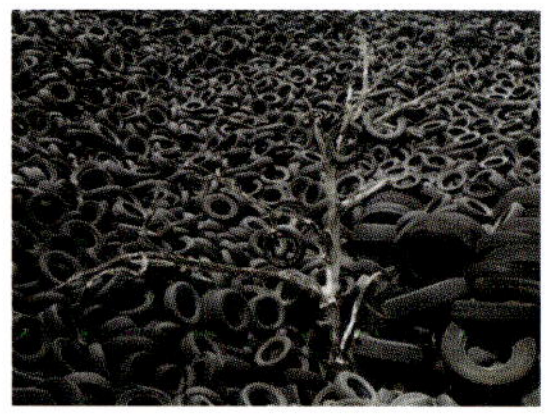

**70 /** Tire dump 1
*France, 2008.*

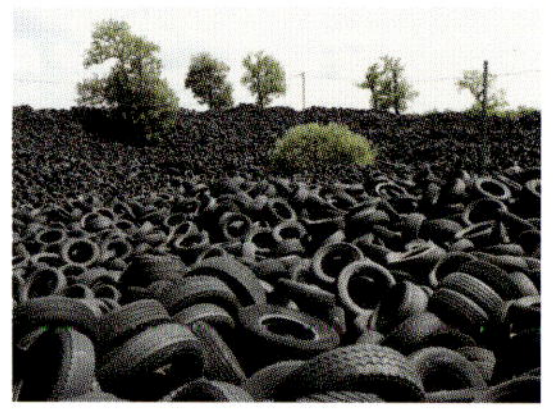

**71 /** Tire dump 2
*France, 2008.*

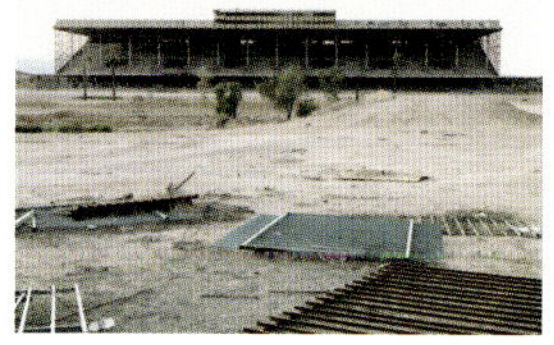

**73 /** Abandoned grandstand
*USA, 2007.*

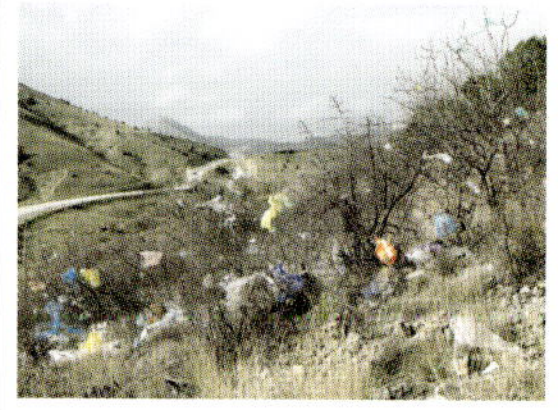

**74 /** Trash from an open-air dump
*Ukraine, 2007.*

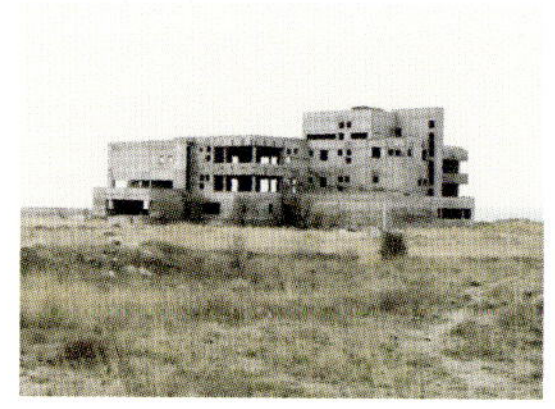

**75 /** Abandoned construction site
*Ukraine, 2007.*

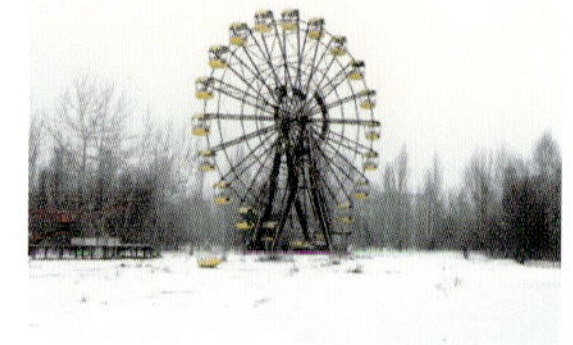

**77 /** Chernobyl 1. The great wheel
*Pripyat, Ukraine, 2007.*

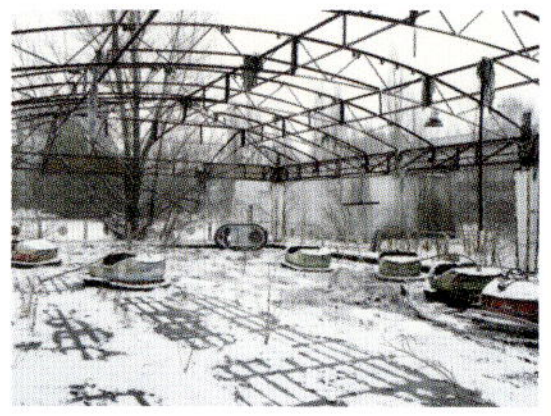

**79 /** Chernobyl 2. Bumper cars
*Pripyat, Ukraine, 2007.*

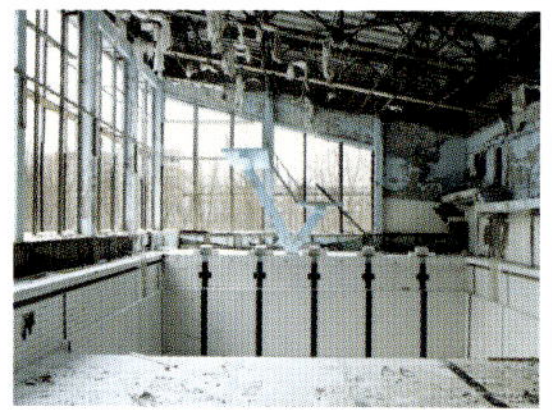

**81 /** Chernobyl 3. Swimming pool
*Pripyat, Ukraine, 2007.*

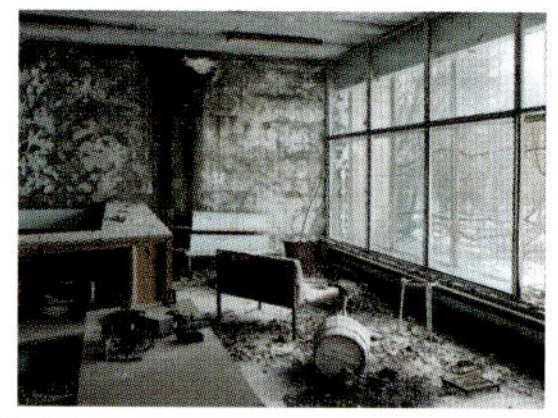

**82 /** Chernobyl 4. Reception room
*Pripyat, Ukraine, 2007.*

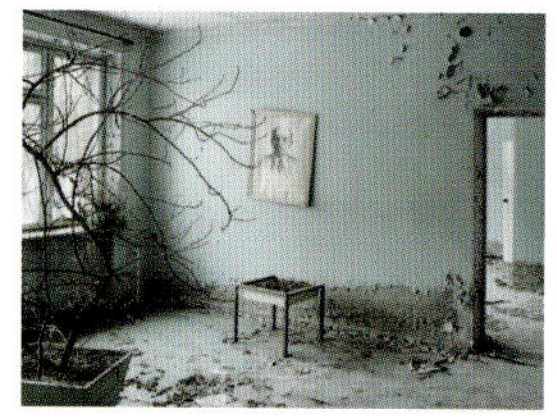

**83 /** Chernobyl 5. Waiting room
*Pripyat, Ukraine, 2007.*

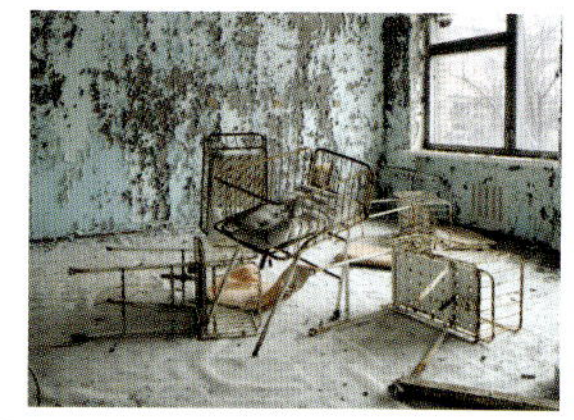

**85 /** Chernobyl 6. Maternity wing
*Pripyat, Ukraine, 2007.*

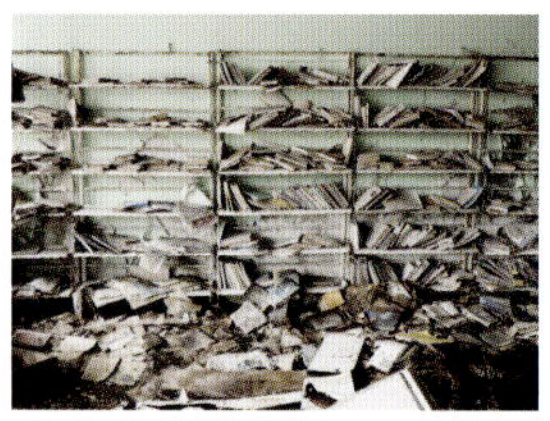

**86 /** Chernobyl 7. School library
*Pripyat, Ukraine, 2007.*

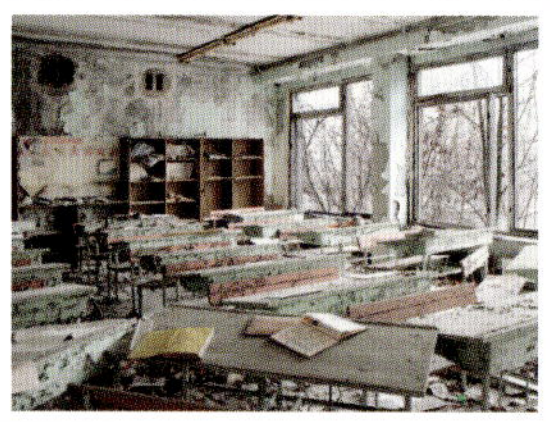

**87 /** Chernobyl 8. Classroom
*Pripyat, Ukraine, 2007.*

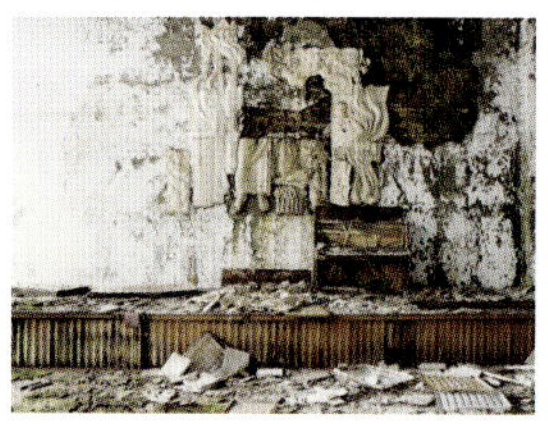

**88 /** Chernobyl 9. Hospital social hall
*Pripyat, Ukraine, 2007.*

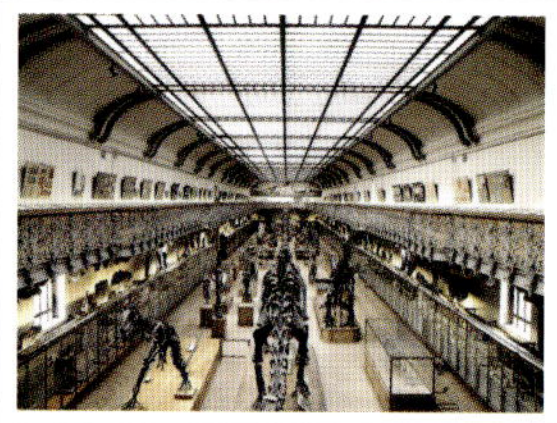

**90 /** Gallery of dinosaurs
*Museum of natural history, France, 2007.*

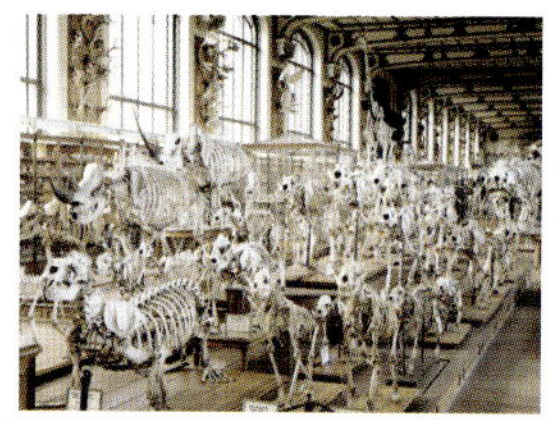

**91 /** Gallery of comparative anatomy 1
*Museum of natural history, France, 2007.*

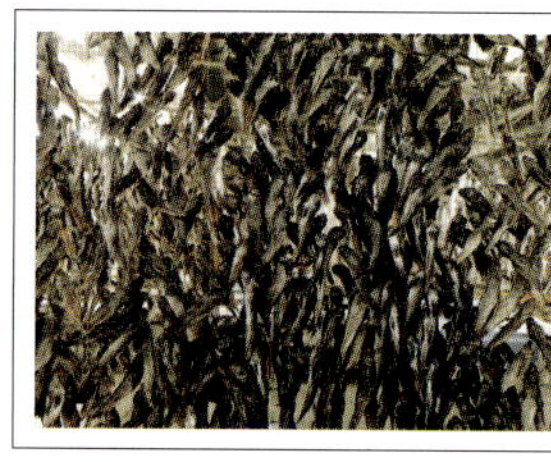

**92 /** Contraband fish
*Crimea, Ukraine, 2007.*

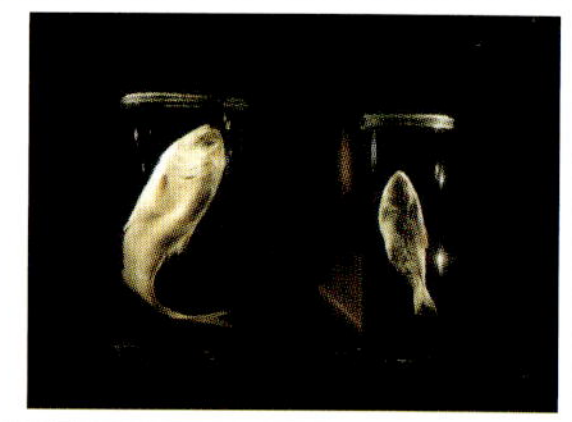

**93 /** Gallery of extinct species
*Museum of natural history, France, 2007.*

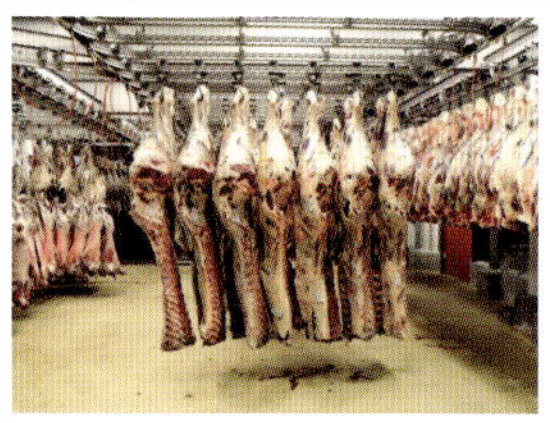

**94 /** The Rungis Market 1
*France, 2007.*

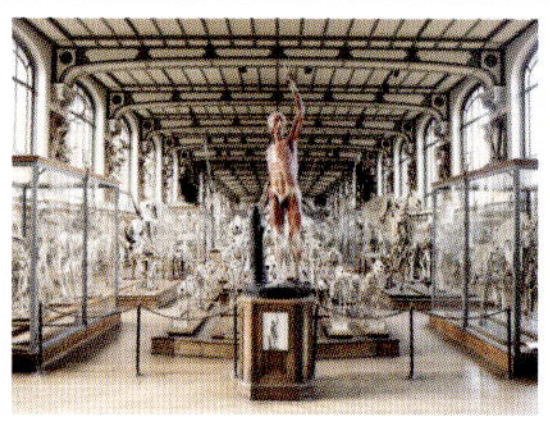

**95 /** Gallery of comparative anatomy 2
*Museum of natural history, France, 2007.*

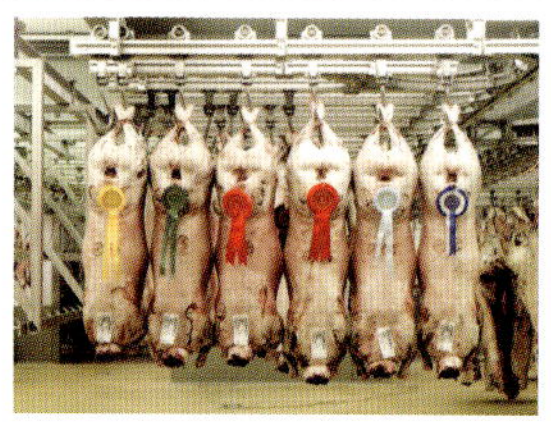

**97 /** The Rungis Market 2
*France, 2007.*

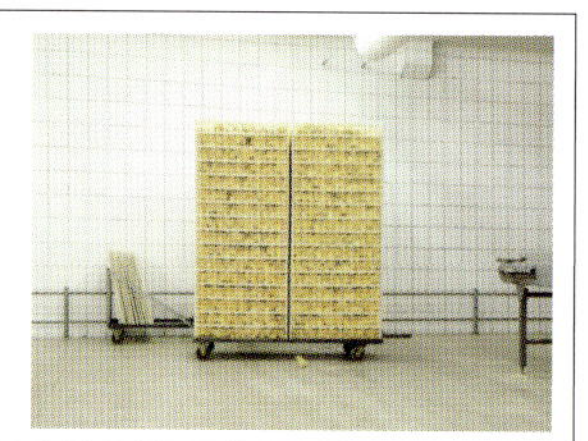

**99 /** Caged chicks
*Ukraine, 2007.*

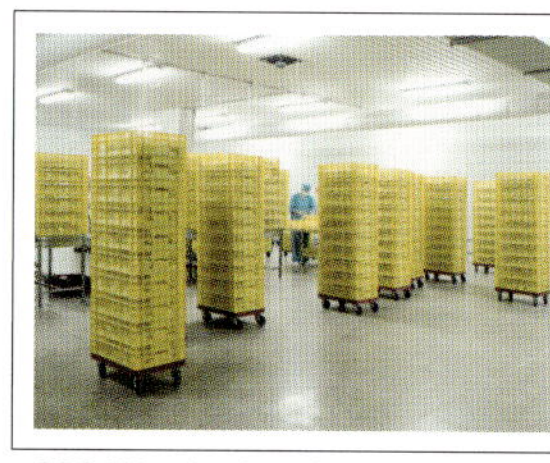

**100 /** Vaccination of chicks
*Ukraine, 2007.*

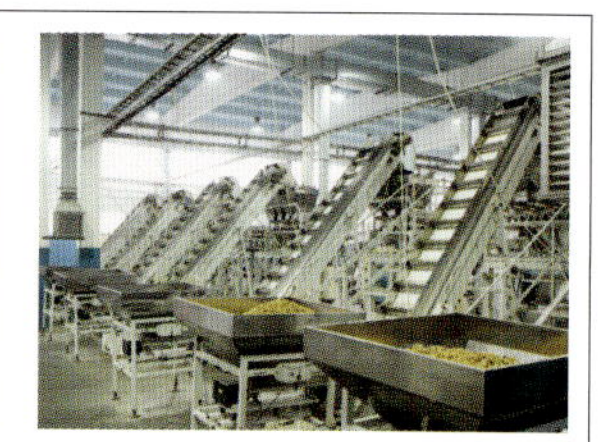

**101 /** Cracker factory
*Ukraine, 2007.*

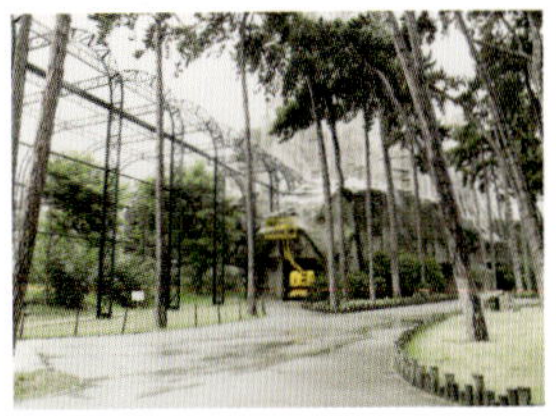

**102 /** Zoo 1
Vincennes, France 2006

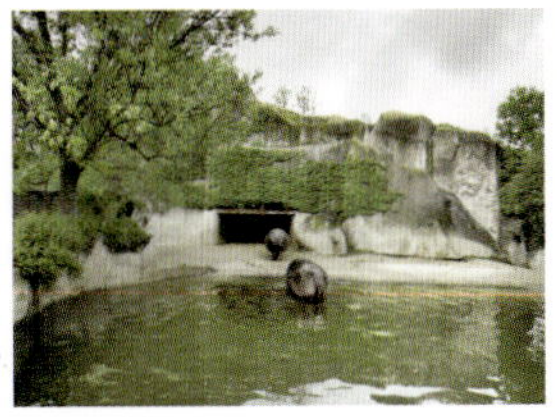

**103 /** Zoo 2
Vincennes, France 2006

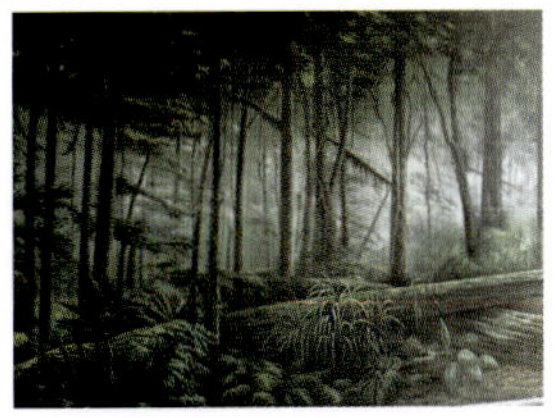

**104 /** Orangutan cage 1
Jardin des Plantes, France 2007

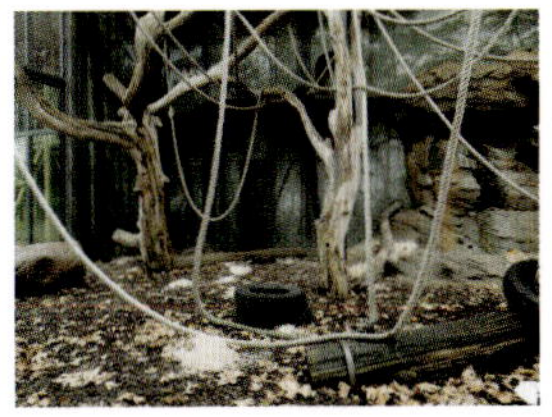

**105 /** Orangutan cage 2
Jardin des Plantes, France 2007

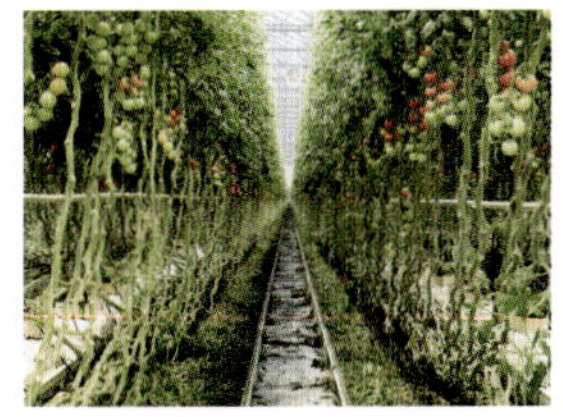

**107 /** Tomato greenhouse 1
Holland 2008

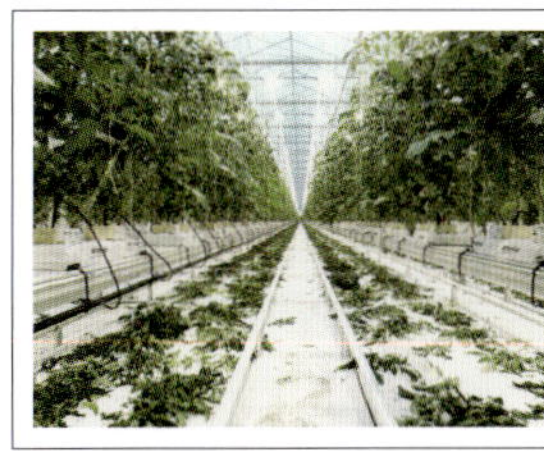

**108 /** Tomato greenhouse 2
Holland 2008

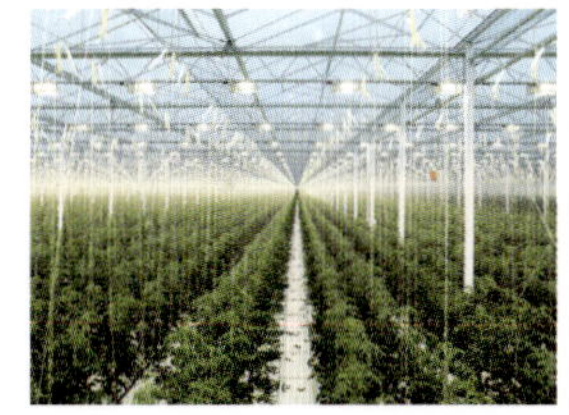

**109 /** Tomato greenhouse 3
Holland 2008

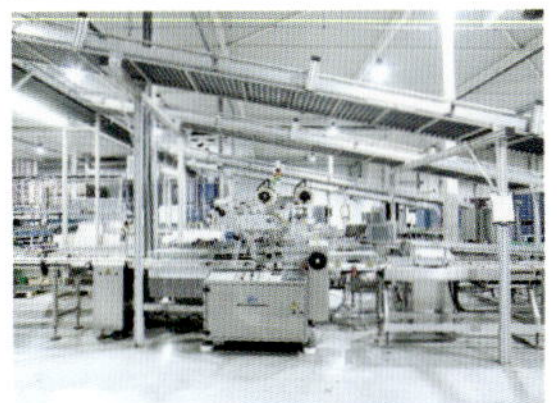

**111 /** Tomato packaging production line 1
Holland 2008

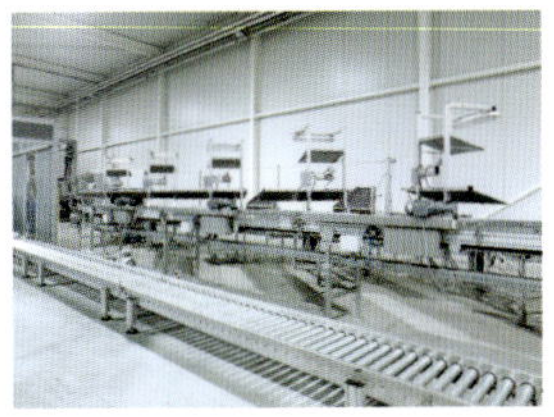

**112 /** Tomato packaging production line 2
Holland 2008

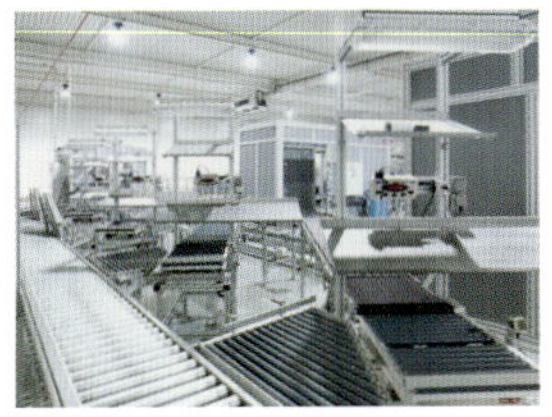

**113 /** Tomato packaging production line 3
Holland 2008

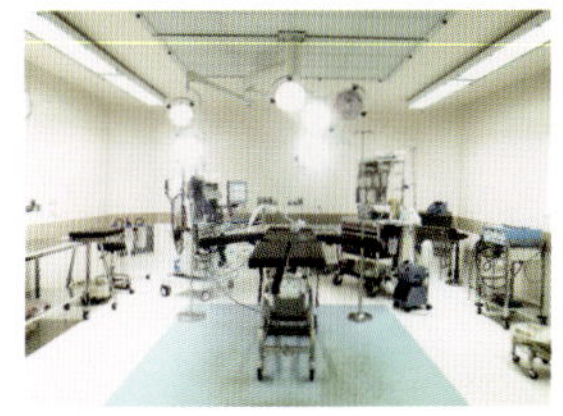

**115 /** Operating room
Saint Camille Hospital, France 2008

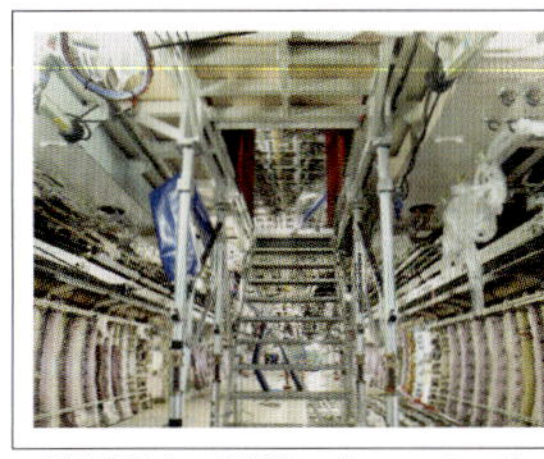

**116 /** Airbus A380 under construction 1
France 2008

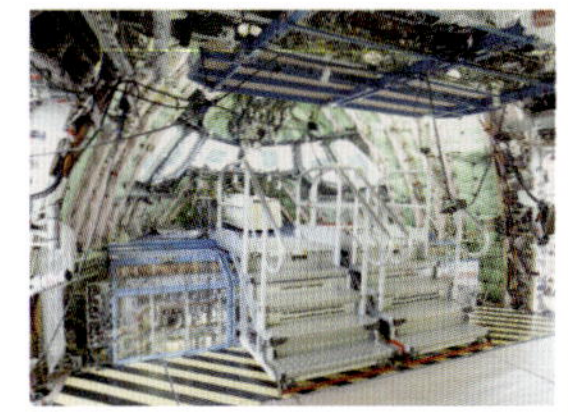

**117 /** Airbus A380 under construction 2
France 2008

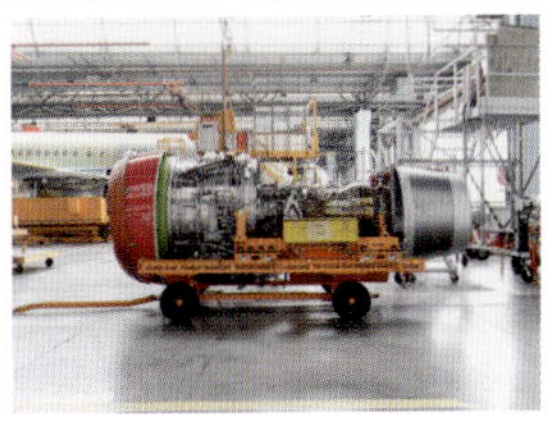

**118 /** Airbus A320 engine
France 2008

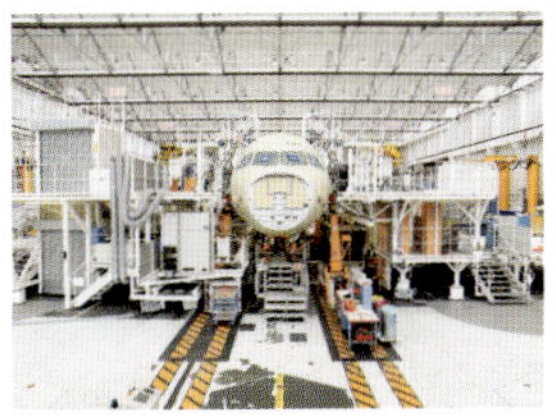

**119 /** Airbus A320 assembly line
France 2008

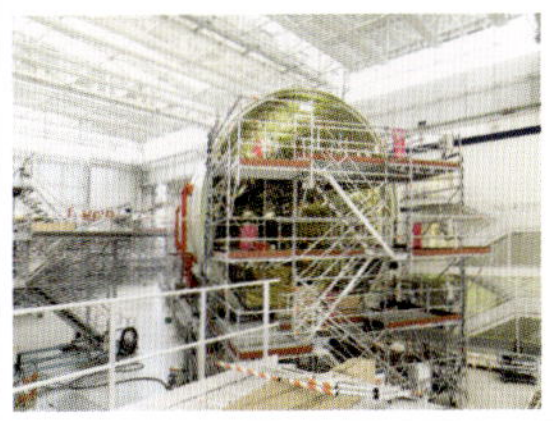

**120 /** Airbus A380 under construction 3
France 2008

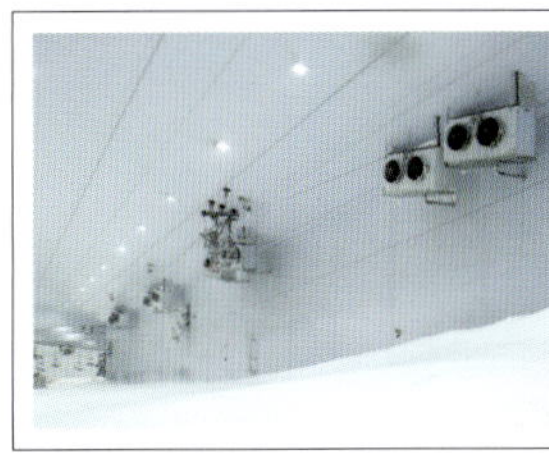

**122 /** Ski Dubai 1
United Arab Emirates 2008

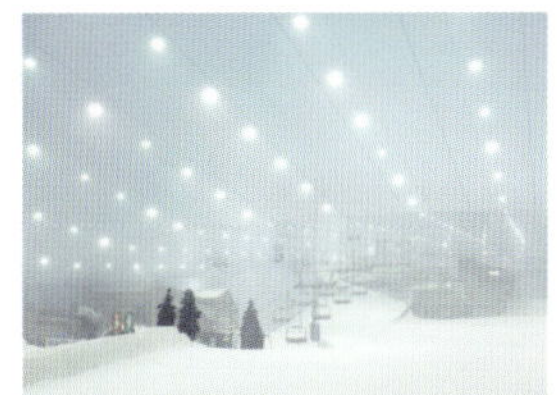

**123 /** Ski Dubai 2
United Arab Emirates 2008

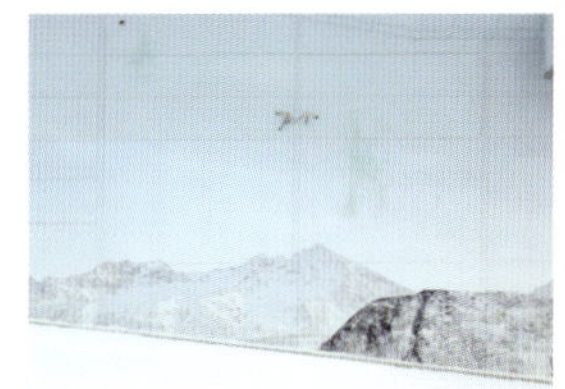

**125 /** Ski Dubai 3
United Arab Emirates 2008

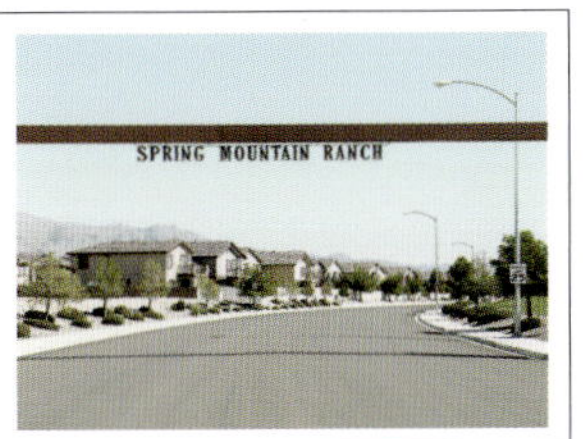


**127 /** Main street
Spring Mountain Ranch, USA 2007

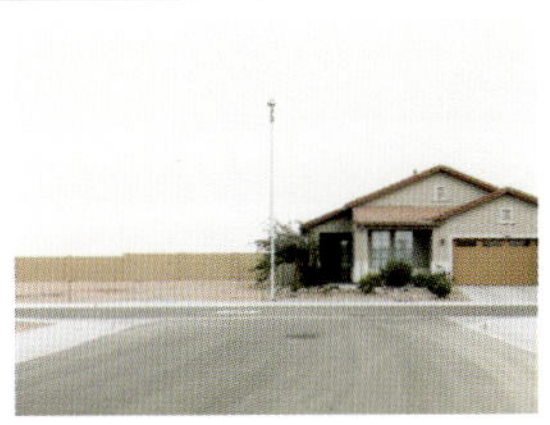

**129 /** Houses under construction
USA 2007

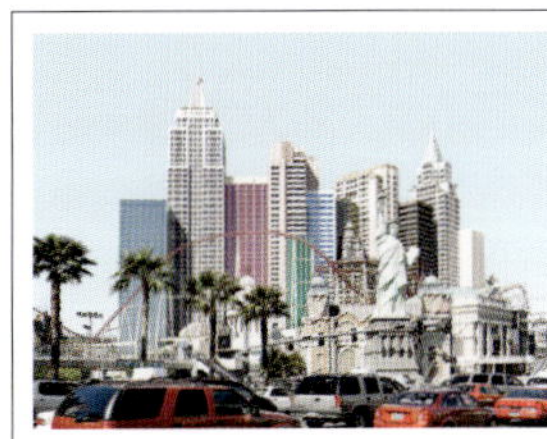

**130 /** Las Vegas casino 1
Nevada 2007

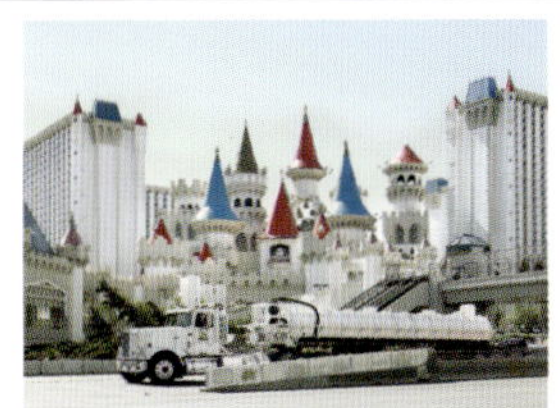

**131 /** Las Vegas casino 2
Nevada 2007

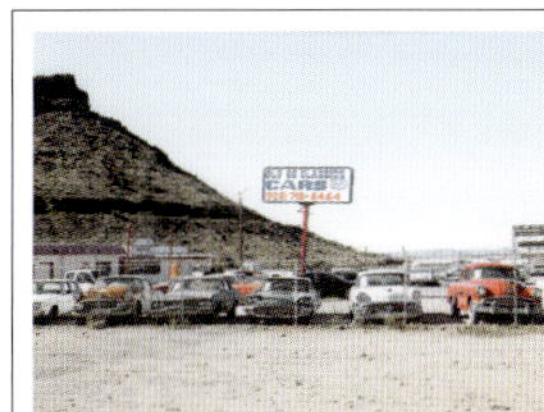

**132 /** Used car lot 1
Route 66, USA 2007

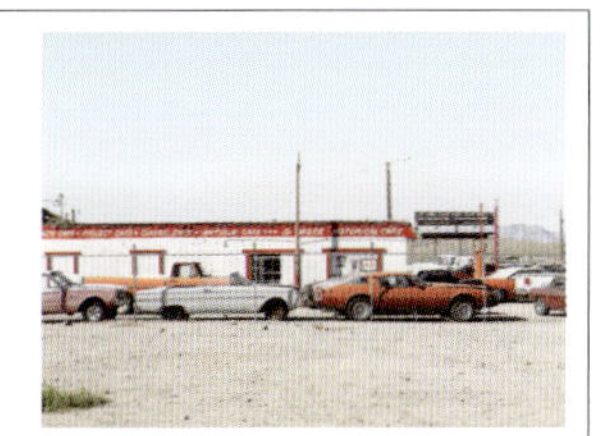

**133 /** Used car lot 2
Route 66, USA 2007

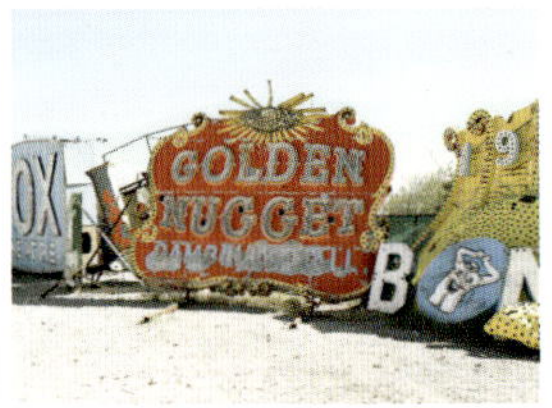
**134 /** Neon sign boneyard 1
*Neon Museum, Las Vegas, Nevada, 2007.*

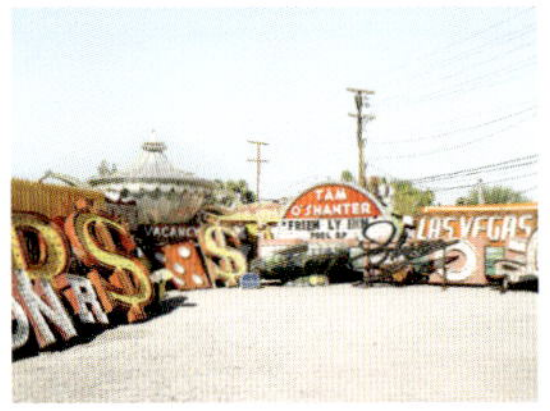
**135 /** Neon sign boneyard 2
*Neon Museum, Las Vegas, Nevada, 2007.*

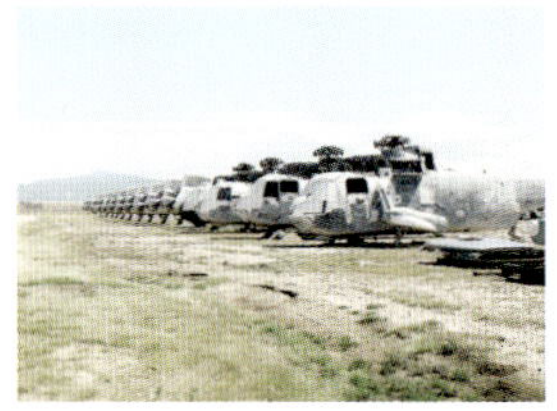
**137 /** Military base 1
*Tucson, Arizona, 2007.*

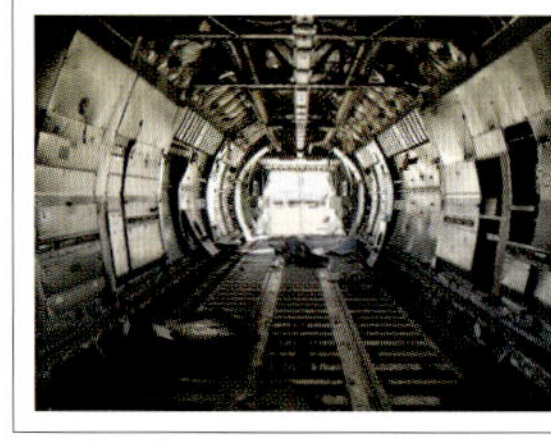
**138 /** Military base 2
*Tucson, Arizona, 2007.*

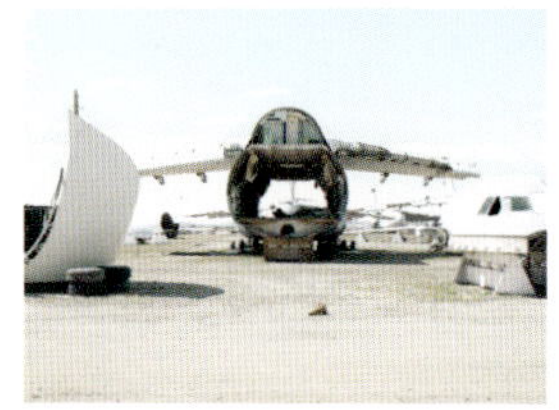
**139 /** Military base 3
*Tucson, Arizona, 2007.*

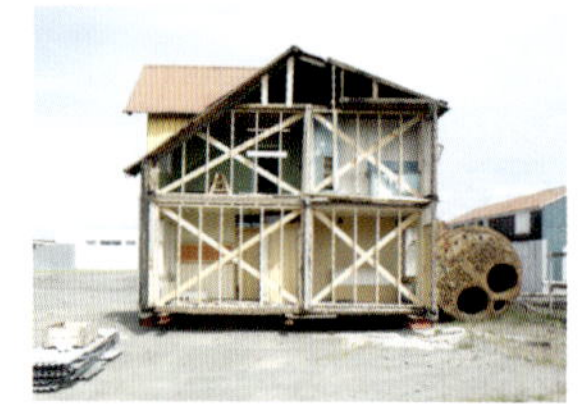
**141 /** House demolition
*Iceland, 2007.*

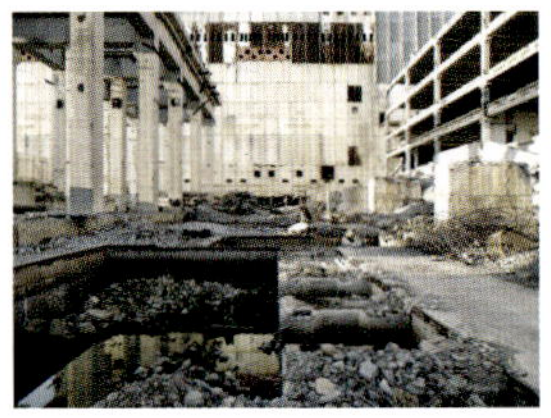
**143 /** Nuclear plant demantlement 1
*Ukraine, 2007.*

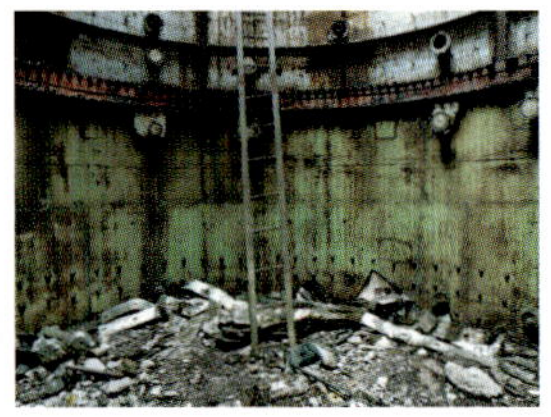
**144 /** Nuclear reactor core
*Ukraine, 2007.*

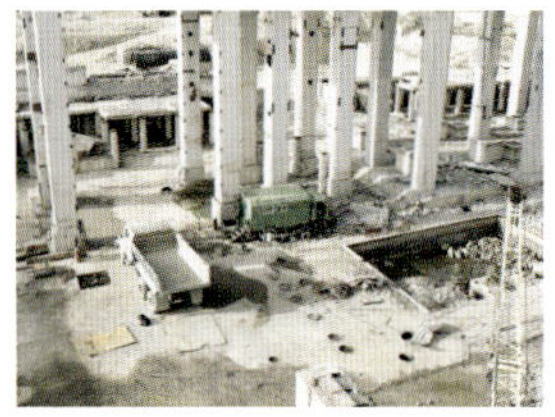
**145 /** Nuclear plant demantlement 2
*Ukraine, 2007.*

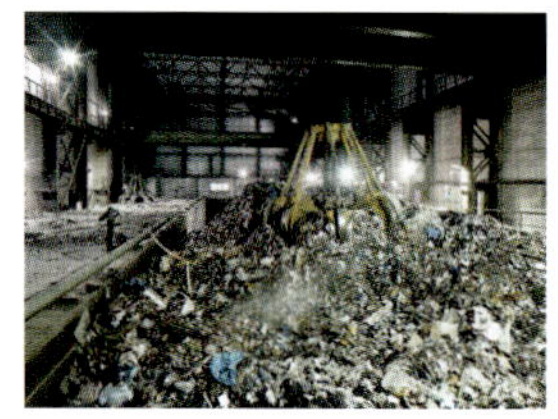
**147 /** Incineration plant 3
*Ukraine, 2007.*

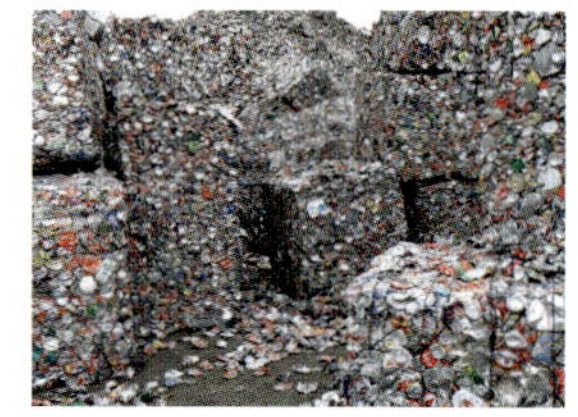
**149 /** Recycling plant 1
*Belgium, 2008.*

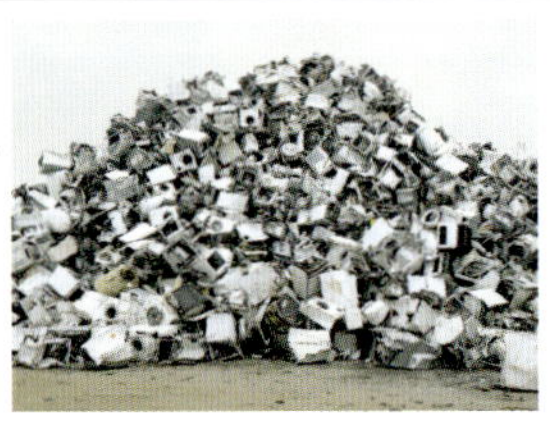
**150 /** Recycling plant 2
*Belgium, 2008.*

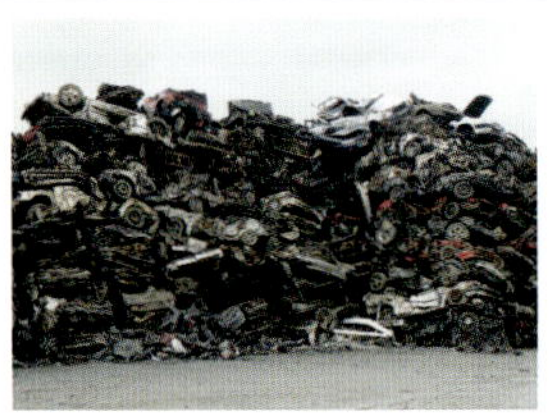
**151 /** Recycling plant 3
*Belgium, 2008.*

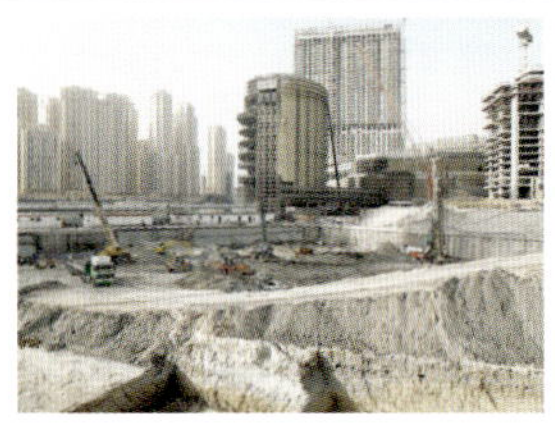
**153 /** Construction site 1
*Dubai, United Arab Emirates, 2008.*

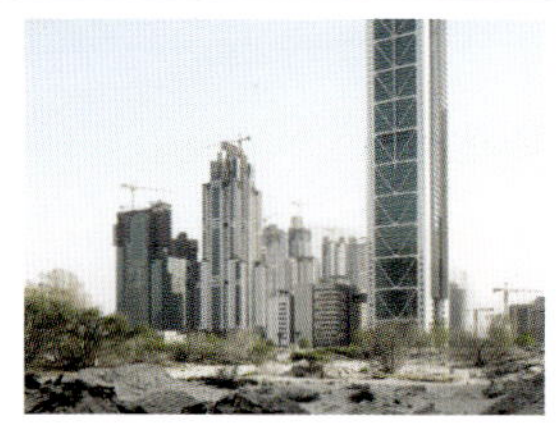
**155 /** Construction site 2
*Dubai, United Arab Emirates, 2008.*

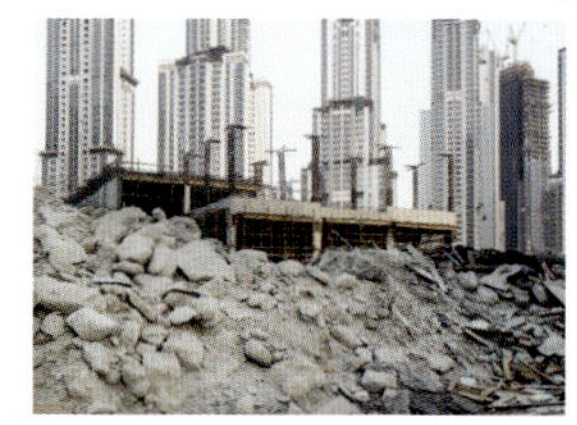
**157 /** Construction site 3
*Dubai, United Arab Emirates, 2008.*

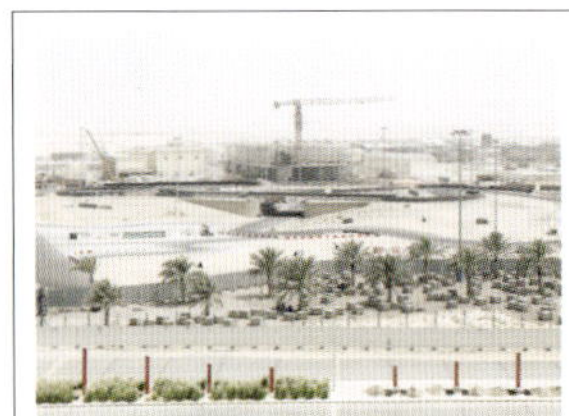
**158 /** Construction site 4
*Dubai, United Arab Emirates, 2008.*

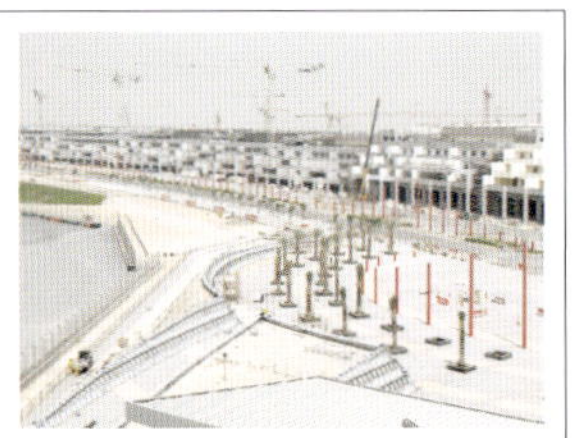
**159 /** Construction site 5
*Dubai, United Arab Emirates, 2008.*

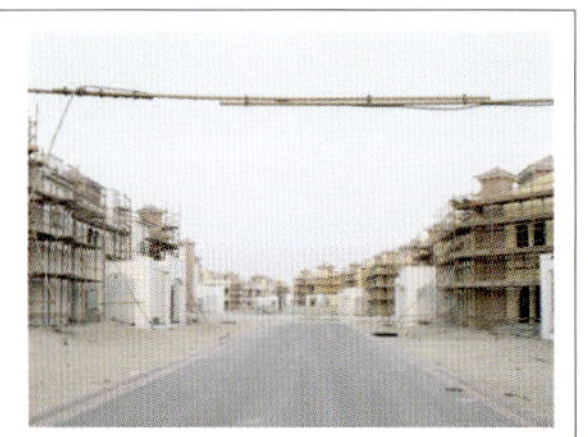
**161 /** Construction site 6
*Dubai, United Arab Emirates, 2008.*

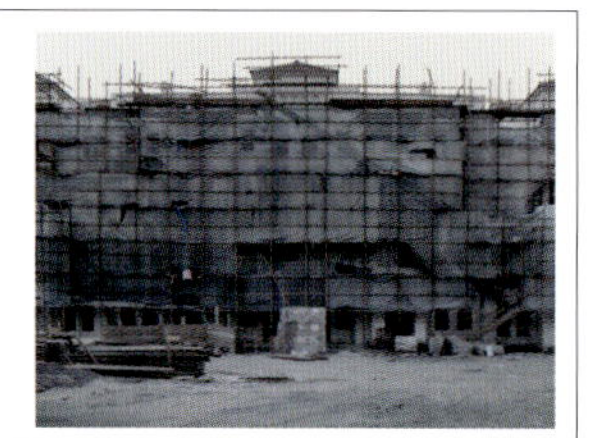
**163 /** Construction site 7
*Dubai, United Arab Emirates, 2008.*

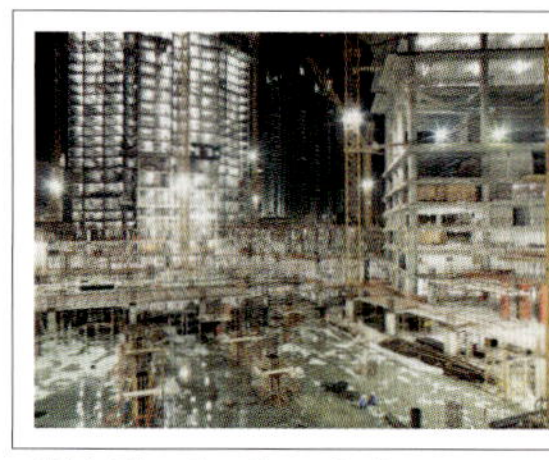
**164 /** Construction site 8
*Dubai, United Arab Emirates, 2008.*

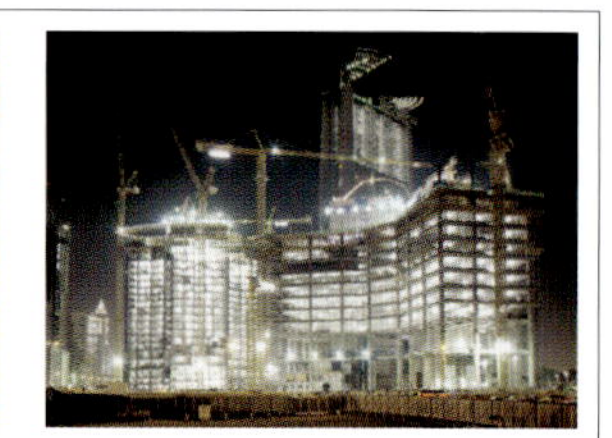
**165 /** Construction site 9
*Dubai, United Arab Emirates, 2008.*

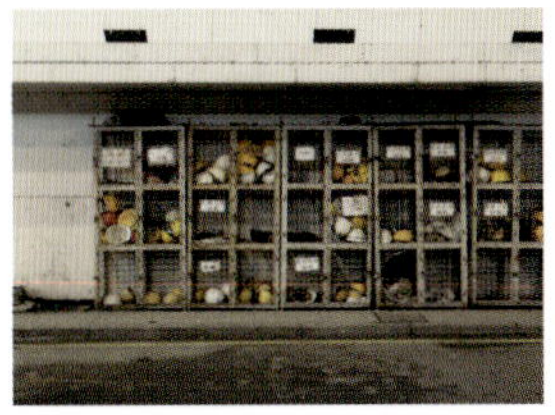

**167 /**
*Macao, China. 2008.*

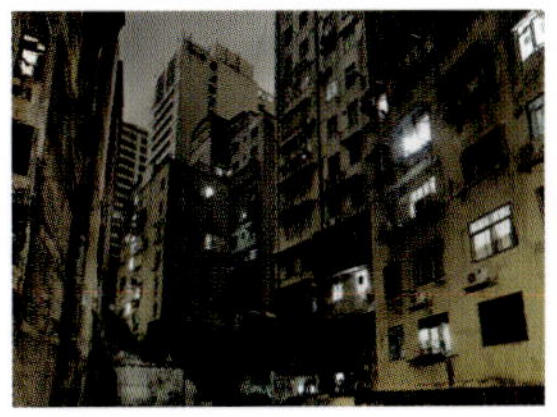

**168 /**
*Macao, China. 2008.*

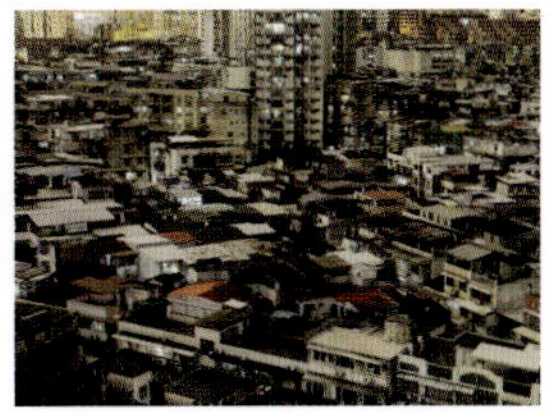

**169 /**
*Macao, China. 2008.*

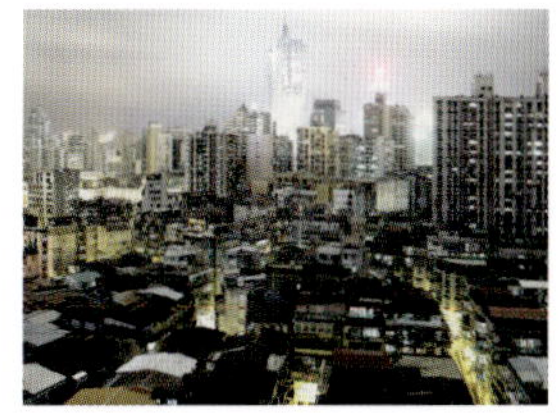

**171 /**
*Macao, China. 2008.*

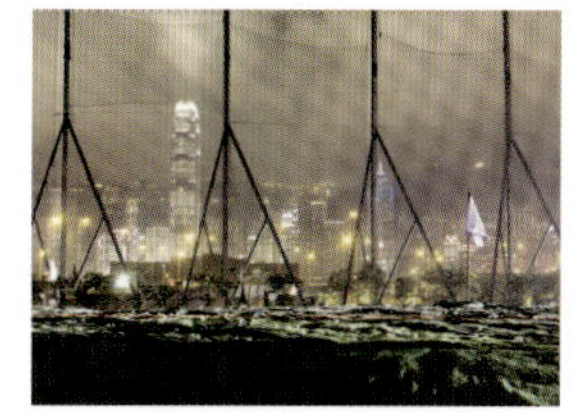

**173 /**
*Hong Kong, China. 2008.*

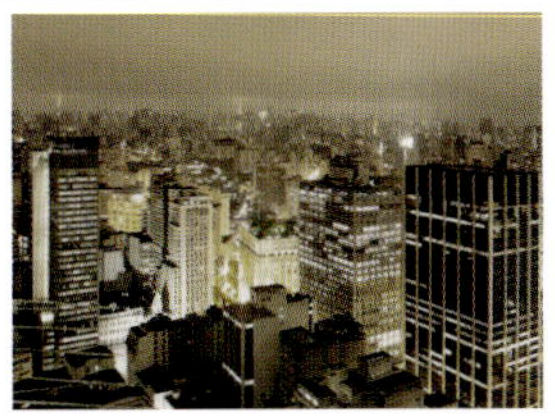

**175 /**
*São Paulo, Brazil. 2008.*

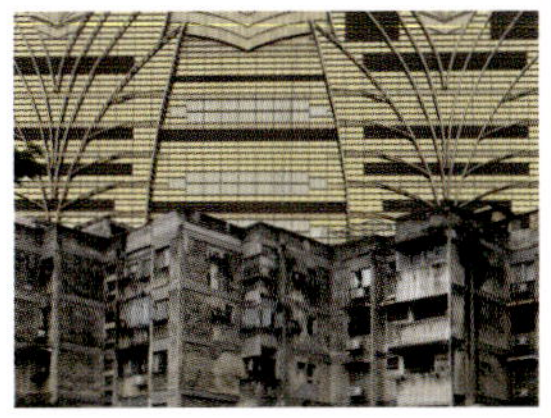

**177 /**
*Macao, China. 2008.*

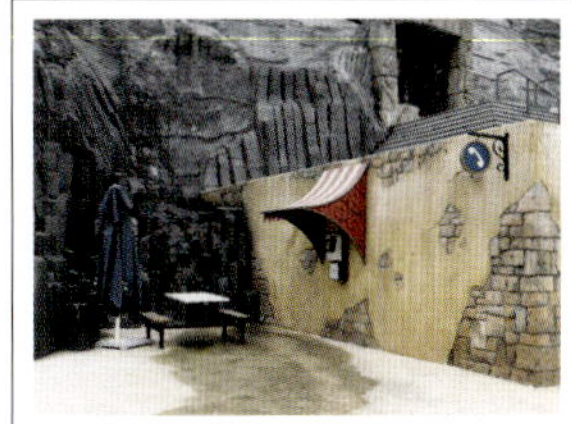

**178 /** Casino entrance
*Macao, China. 2008.*

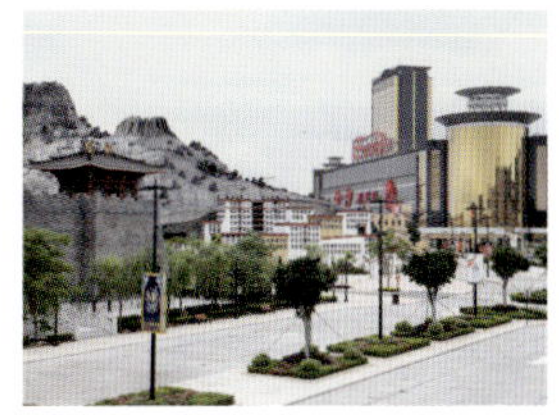

**179 /** Casino
*Macao, China. 2008.*

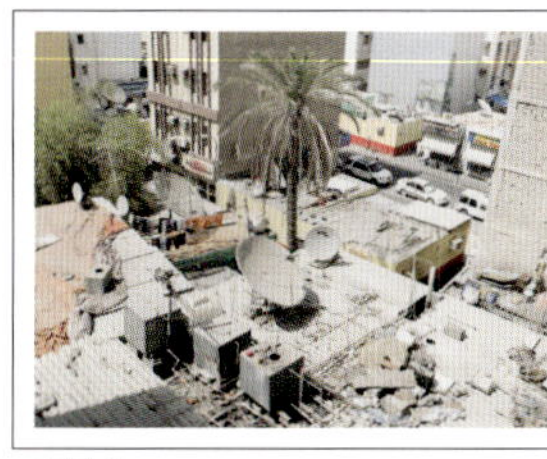

**180 /**
*United Arab Emirates. 2008.*

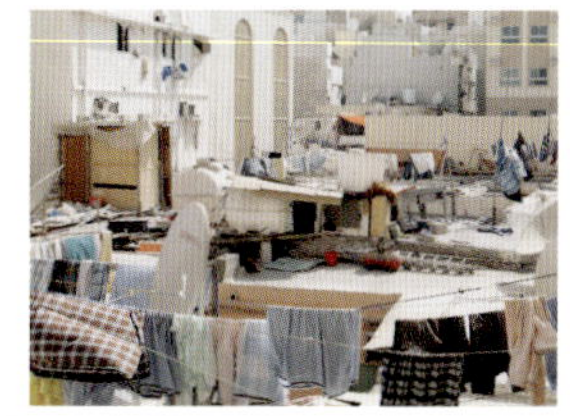

**181 /**
*Dubai, United Arab Emirates. 2008.*

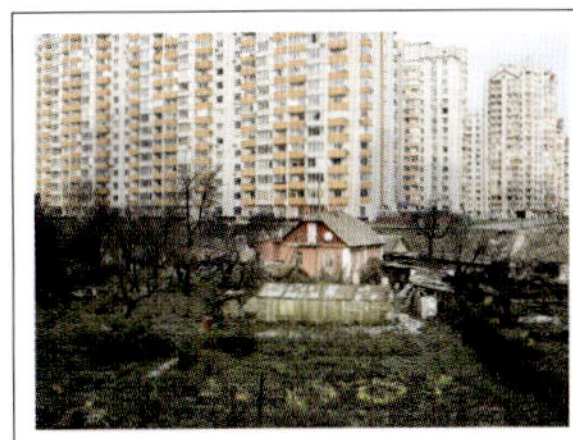

**182 /**
*Kiev, Ukraine. 2007.*

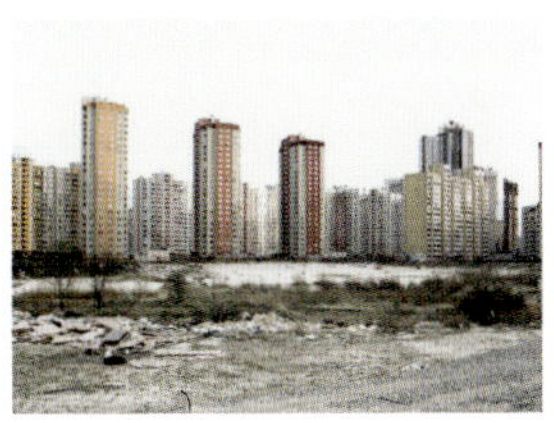

**183 /**
*Kiev, Ukraine. 2007.*

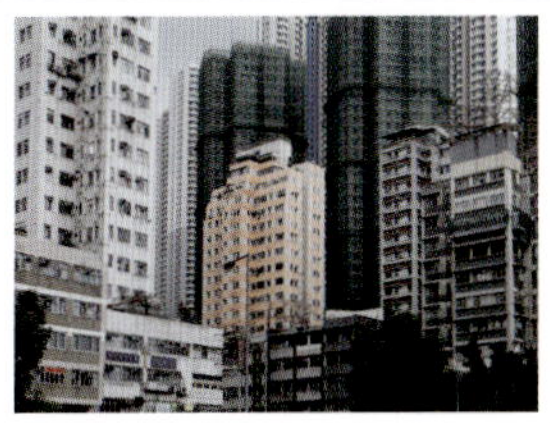

**185 /**
*Hong Kong, China. 2008.*

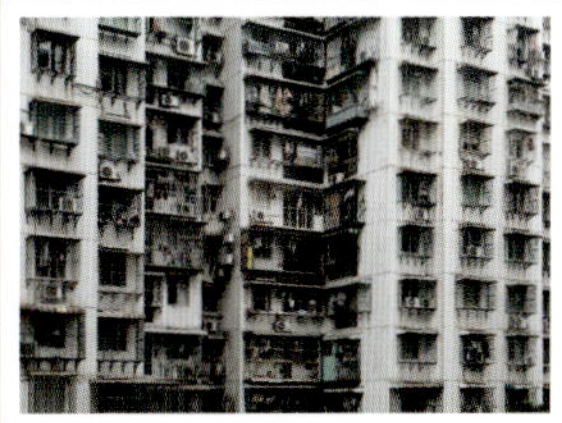

**186 /**
*Macao, China. 2008.*

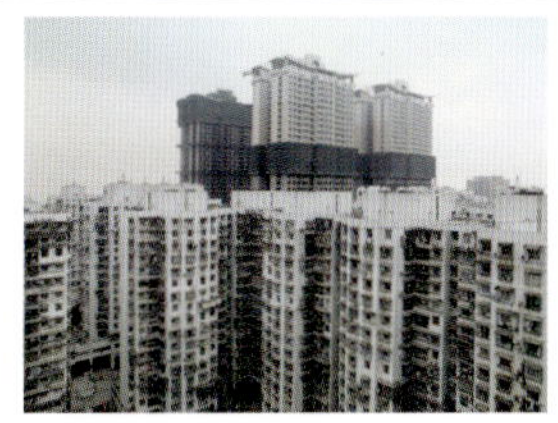

**187 /**
*Macao, China. 2008.*

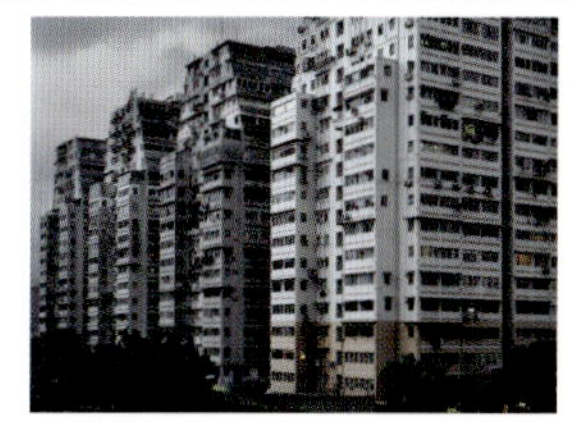

**189 /**
*Hong Kong, China. 2008.*

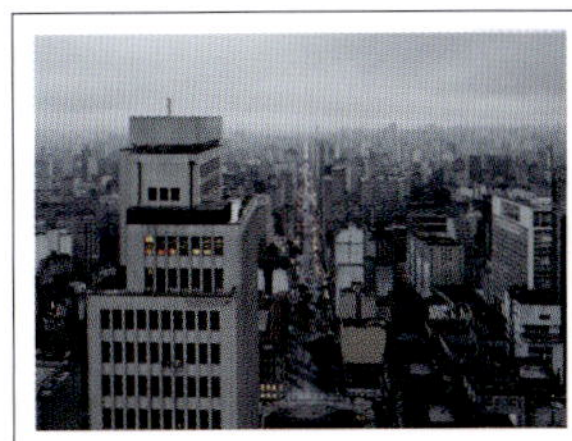

**190 /**
*São Paulo, Brazil. 2008.*

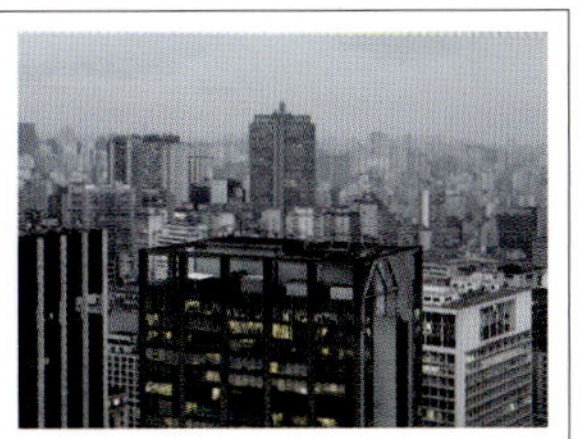

**191 /**
*São Paulo, Brazil. 2008.*

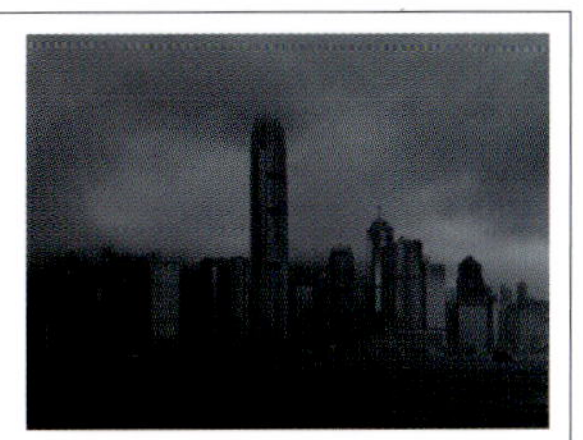

**193 /**
*Hong Kong, China. 2008.*

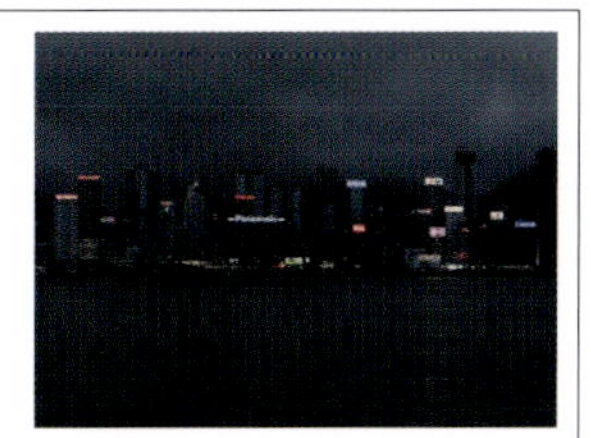

**195 /**
*Hong Kong, China. 2008.*

Here to Stay Project **Olivier Bourgeois**

# "Never has the conflict between the human race and the environment upon which its existence depends reached such a critical point."

At a press conference in Tokyo on March 30, 2005, Hans Van Ginkel, adjunct secretary general of the United Nations, cited the conclusions of the UN report the *Millennium Ecosystem Assessment:* "About 60% of the ecosystems that permit life on earth have been damaged . . . Human activity exercises such pressure on the natural functioning of the planet that the capacity of the ecosystems to respond to the demands of future generations can no longer be taken for granted."

The inevitable exhaustion of natural resources, the abnormally high extinction rate of animal and vegetable species, and global warming have violently revealed the fragility of our future as well as of our civilization. Our scientific knowledge and technological mastery have never reached such a high point, but we are on the point of being trapped by our most elementary needs: pure air to breathe, water to drink, soil to grow our food, a stable climate so we can live in security.

Never has the conflict between the human race and the environment upon which its existence depends reached such a critical point. On one side, we have progress and its growing appetite to feed urban demands, on the other a biodiversity that is drying up progressively and a planet that is being asphyxiated by our own inventions. Is a tragedy being written? Can we reverse the course of things? Do we want to?

It is in this context that the film and photography project *Nous Resterons sur Terre/Here to Stay* were conceived; out of the will to take a careful look at Earth both as our awesome heritage, and as an object on which a civilization remains persuaded that it can construct imaginary and limitless cities.

For long months, our gazes crossed, intermingled . . . What good luck to have shared this project with you, Cédrtic, to have felt this fascination and this fear together in the face of often unreal settings, to have told the same story from different angles, to have taken the time to contemplate the world.

What remains of this project? I don't know. I do know, on the other hand, what remains with me: the intensity of the moments we lived together, the emotion that seizes me as I look at these photos, a sincere friendship. And one certainty: we're here to stay.

## Cédric Delsaux interview **Jean-Sébastien Stehli**

Cédric Delsaux covered the planet for more than a year with the team of the documentary *Nous Resterons sur Terre/Here to Stay*, but traveling in his own right as photographer and artist. Result: a book unlike any other. These photos demonstrate the impact of humankind on the planet, but Cédric Delsaux has found a completely new way of looking at our world as it appears after centuries of human presence. His work is neither an accusation nor a picture postcard. He is both tender and perceptive.

**How was your book conceived?**
*A Common Destiny* grew out of the documentary film *Nous Resterons sur Terre*. The directors proposed that I accompany the crew, not to photograph the filming but to express my own vision of the places they were visiting. So I chose five or six places that were emblematic of the westernization of the world. I didn't know what I was going to see. I went on site with no preconceived ideas. I knew one thing, however: I wanted to do more than create a factual record, because everyone knows the state of the planet. I did not want to restrict myself to presenting a cold, distant view. I wanted to get involved. My intention was to tell the story of the ties I have made with this world. I am filled with admiration, hatred, fascination, and disgust all at once.

***A Common Destiny* is neither the book of a photojournalist nor the book of an environmentalist. Where do you situate yourself?**
The environmentalist puts himself in the planet's position. He photographs panoramas, landscapes. I don't speak for the planet; my place is as an individual looking at what we've done. Not once have I gone up in a helicopter. I want to be at the level of the pedestrian, looking at the world from eye level. This approach allows me to retain my contact with it. My photography, as such, is more a mirror than a window.

**Curiously, no people appear in your photos . . .**
That's true. They are either tiny or they are phantoms, because I generally work with a very long time exposure. My intention isn't to tell about the conditions of the Indians or Chinese construction workers in Dubai, for example. You could say that my photos are post-apocalyptic: they are the negatives taken after man has left the scene. We've disappeared and I photograph what we leave behind us. In these photos, about the only living being is one chick that escaped from a dolly full of stacked cages. The humans you see are always in lab coats. They are working in an environment that reduces them to the status of executioners, or worse, of robots.

Jean-Sébastien Stehli has been a journalist for thirty years. First at *Le Point* and later at *L'Express*, where he was editor in chief of the society section and later of arts and entertainment. He was a longtime correspondent in the U.S., and founded the monthly magazine *Ici New York (New York Day by Day)*. He is currently a managing director at Agence Andrée Putman, an interior design and architecture agency, and an avid collector of photography.

**What's very curious, in your images that show the devastation that we are responsible for, is that they elicit a great sweetness, even when you are photographing a machine swallowing a mountain, or even mountains of wrecked cars . . .**

I try to put a maximum of humanity into something that is no longer human, and which is nevertheless the world we have made. The contradiction is evident. We have a tendency to reassure ourselves with a Manichaean vision of the world, those who are good, those who do good, and then everyone else. Whatever is dazzling, whatever is somber, what is considered a beautiful landscape or an ugly one . . . Well, I can only challenge that, because for me all these notions overlap, constantly mix together. This is why I can't stop myself from seeing beauty in the ugly. And vice versa. Once this idea has been postulated, I can no longer protest, teach lessons in morality. I have to learn to live with all this without violence and without hypocrisy: gently.

**How?**

Partly by working with color. The more I advance, the less contrast my images have. I work on the shading that comes only with color. When I shoot cars, for example, I need to touch them. I want to pay homage to the materials. And when I photograph an enormous machine in a manganese quarry, rather than treat it as an aggressive structure, I work on its halftones. I'm a little like someone who returns to photograph the world after the end, who has a tender feeling for what is no longer there.

**There are two ways of awakening people's consciences about the world: either show its beauty, or, on the contrary, expose its horrors. You've chosen yet a third way.**

I'm not given to apology, or to denunciation. If you want make a huge production of it, you show the extent of the horror spread over hundreds of kilometers. I'm the opposite: I limit the framing of the photo. I try to keep it on a human level. If your frame is too big, it says too much, or too little, about the subject. You have to find the right distance: a distance between that of the photojournalist at the heart of the action and that of the plastic surgeon who removes himself from the world; the distance, in actuality, is what invites introspection.

**Even a poetic photo, like the hut in the Amazonian forest with its parabolic antenna like a delicate flower, is the sign of a disaster foretold.**

Yes. Exactly a week after my photo, the village was connected by Internet to the rest of the world, and this antenna, so fragile, almost beautiful, finally dragged behind it the slow destruction of a culture and of the forest.

**The photo of Hong Kong's bay has an end-of-the-world murkiness . . .**

But it produces a sad sweetness. It's a darkness that is almost light, it's the idea of a world that is consuming itself, almost without knowing it, out of general indifference. Someone is surely going to tell me that Hong Kong is scarcely on the point of disappearing . . .

**Yes, that's manipulation!**

But I have the right to do it because I am not a photojournalist! It's obviously just one more storm over the city at nightfall, but placed in such a series as this, it achieves impressive proportions. All my photos are, effectively, reinterpretations. I go looking for primary material in reality that I can take another look at it in my own way. The last image in the book, Hong Kong's bay again, has the same dynamic: I see in it the madness of the world, disappearing behind neon advertising signs. And since I myself have worked for some of these companies, I don't presume to teach any lessons. This is more of a work about the way we perceive things. I believe that the real exists only in our perception of it. As a result, to label it or even to protest against it can only be, in my opinion, a misinterpretation.

## Wangari Maathaï

"Ignorance can no longer be an excuse."

The good news is that more and more people throughout the world are becoming aware of the need to treat the Earth as a very precious planet that has to be cared for. The bad news is that there are still many people who see the Earth's resources as something to exploit, to own, to privatize for their own personal gain. They don't seem to realize that these resources are limited and can be depleted to such a degree that our own lives are threatened.

The challenge for us all is the following: either we change our relationship with the planet and with the other forms of life that live on it, or we continue down the path of exploitation, privatization, ownership, and consumption, undermining the ability of future generations to survive on Earth. Do we care about our children's future?

Across the whole African continent, most individuals are still, of necessity, very close to nature. They are directly dependent upon it. In many cases, they still have to draw their water from rivers, wells, and streams, produce their own food from the land they live on, and use wood as their main source of energy. But here, too, urbanization is at work.

Urbanization is a process that very rapidly disconnects us from nature. We are beginning to live in what we call "concrete jungles." We are settling in this developed world where everything is controlled by science and by knowledge that is largely unfamiliar to the inhabitants of rural areas. Water comes from a pipe and food from the supermarket. Travel is provided by such devices as trains, buses, and autos. We live and work in what they would call an artificial world. Those people who live in rural areas and who are still closely connected to the natural world are judged, and therefore judge themselves, to be primitive and underdeveloped, with the result that they, too, end up wanting to be part of this urbanized, technological world.

This is where we lose the connection, the tie, and become strangers to nature. For several years now, I have been working with people in rural areas, trying to encourage them to take care of their land, prevent soil erosion by planting trees, protect rivers and forests, and collect rainwater so that they can grow their own food. Then I go to the urban centers. There, I try to tell the country's parliamentarians, ministers, and decision makers, "You must reconnect yourselves to the natural world, work closely with the people who still live on the land. You must protect the forests and mountains, because if you don't, the taps will soon run dry, the supermarkets will stand empty, and even the air we breathe, laden with pollutants from our technological world, will asphyxiate us."

But guess what? People don't react! They just go on as before, as though they will wait until there really isn't any more water in the pipes or food in the supermarkets, and when the air they breathe makes them sick. It's shameful! It's unbelievable that we humans should behave like that! We don't look to tomorrow. We are too preoccupied with today. Tomorrow seems far away. We make most of our decisions in our own interests, or those of our company, our village or, at best, our country. Not of those who will come after us in the world. This way of going about things is incoherent and very dangerous. We need to change our way of thinking, embrace reality as a whole, and really begin to regard the Earth as our home!

The people sounding the alarm are scientists or activists concerned with these specific issues. They are doing it to get politicians and corporations to make decisions. But politicians have other things on their mind: elections, power, and influence! They are obsessed not only with hanging onto their national positions, but also with competition beyond their own borders, sometimes sacrificing the environment in the process.

These conflicts of interest are an obstacle to breaking the environmental deadlock. There's no point holding Earth summits and international conferences, making all sorts of declarations, presenting alarming reports and setting up a United Nations environmental program if we cannot establish a common agenda for action and then abide by it.

In addition, certain developing countries fear that the industrialized nations, already very powerful, both politically and economically, will use the environment as a pretext for subjugating them. Because adopting developed countries' technologies, for example to reduce gas emissions, will probably slow their pace of development.

How can we address this problem? I don't know. We have reached a certain degree of indifference to the common good. Despite all we may say, actions speak louder than words, and these show that we are most often motivated by covetousness, selfishness, and the desire to stay ahead of others, to disregard them completely. Mahatma Gandhi made this lucid comment: "The Earth provides enough to satisfy every man's need, but not every man's greed." What can be done to create a world that thinks of itself as a unified world, a human family? I don't know. It hasn't happened yet. We

Wangari Muta Maathaï was born in Kenya in 1940. Recipient of a scholarship given by the Kennedy Administration, she obtained a bachelor's degree in biology at Mount St. Scholastica College in Atchison, Kansas, in 1964. She continued with graduate studies in Pittsburgh, then in Munich, Germany. She returned to Kenya, enrolling in the School of Veterinary Medicine in Nairobi, where she earned her doctorate in 1971. Maathaï created the Green Belt Movement, which has assisted women with planting over 40 million trees worldwide. She was elected to Kenya's parliament in 2002, and became Assistant Minister for the Environment in 2003. She received the Nobel Prize for Peace in 2004 for her work on behalf of sustainable development, for democracy, and for peace. She is the first African woman in history to be honored with this distinction.

haven't yet raised our consciousness to a sufficient level. We talk about it, but our actions contradict our words. When you hear governments hold forth in international forums and then watch what they actually do, the difference is staggering.

And we still haven't tackled the issue of poverty! A subject that has been endlessly debated, poverty also engenders environmental degradation. It was the Prime Minister of India, Indira Gandhi, who first established a connection between poverty and the environment, in Stockholm in 1972.

If only we were prepared to share our resources more equitably, even if only at a national level! In Kenya, we have experienced political and humanitarian crises, where the poor killed each other, pushed each other aside, because our leaders couldn't agree about the results of the presidential election. In any country, large or small, the absence of a democratic government enabling people to enjoy their own wealth leads to conflict. We must no longer allow poverty and the environment to be set against each other.

The future of a nation is like an African three-legged stool. The carpenter carves it from a single piece of wood, and by the time he has finished, he has created three sturdy legs and a fine base on which one can sit comfortably. The first leg is that of good governance, which requires the respect of human rights and fair distribution of wealth. The second leg is that of the responsible and sustainable management of resources. The third is peace. The seat is the space on which development can take place. Poverty for me, therefore, is a symptom of the failure of a country's three pillars.

But why wait to act until the politicians have woken up? Each one of us must have the moral courage to do what needs to be done. Self-education is extremely important, particularly for those who live in countries where information circulates freely. All too often, we act out of ignorance, and that can no longer be an excuse, for nature has its own laws and doesn't care if we are ignorant. Whether we treat the planet badly or simply don't take care of it, the consequences will be the same! It is therefore in our own interests to educate ourselves, to understand the impact of our collective and individual activities on the environment. And to change.

Each one of us needs a reminder of the things we can do. My Green Belt Movement organizes seminars in which we explain to people just how their deeds destroy the environment and how they can remedy the situation. One of the main activities we have been encouraging for more than thirty years, particularly among rural populations, is tree planting and soil protection. We want to convince people of the need to protect their forests and improve their immediate environment before they even start to think globally. In highly industrialized countries, in order to gain legitimacy and find the right words to awaken people's consciousness, you have to draw on ancestral culture. In Japan, for example, we worked with the Minister of the Environment, Yuriko Koike, to create a message which combined two very ancient traditional concepts: *mottainai,* which means "Let us be grateful for the resources we have, not waste them, and use them respectfully," and *furoshiki,* which is a traditional cloth used to wrap gifts. The alliance of these two customs gave rise to a very powerful public awareness campaign in Japan, also known as the 3R: "Reduce, Reuse, Recycle."

There are so many things we can do! We must be creative and ask ourselves what we can do in our daily lives to help the planet. Some will plant trees and others will avoid chopping them down; some people will protect the soil and others the mountains, where we know the glaciers are melting. Other people still will promote the use of public transport or travel around on bicycles to limit greenhouse gas emissions.

Intensive farming, which was the logical follow-up to the industrialization of society, has played a major role in breaking our bond with nature. Moreover, in the last couple of decades, the monopolization of seed and crop production by a few big businesses has accelerated. Seed varieties are becoming industrial patents, duly filed and protected. The entire worldwide food chain, from production to distribution, will soon be controlled by a handful of companies. In this context, the vulnerability of farmers and communities, particularly in developing regions, is a major problem. When you deal with businesses, you are no longer dealing with individual people, but with entities in league with governments. If we let things go on the way they are, the future will be ripe with conflicts.

It is quite possible to return to a more reasonable form of agriculture. Many individuals and communities are trying to recover their ability to produce their own food. The organic movement is developing extremely rapidly in many parts of the world. It's a very positive development, signaling an awakening of our consciousness. We must absolutely discourage people from eating food that has been grown a long way away, which requires a lot of fuel for delivery. We must go back to eating seasonal produce, grown locally and in an environmentally friendly way.

It is urgent, indeed vital, for us to refuse to allow our way of thinking to be controlled by yet another industry, namely the information industry. We must inform ourselves in a critical manner, not a superficial one; we must educate our children. And we must rediscover the joy of being at one with nature. Go into your local communities, get involved, protect your rivers and streams, combat soil erosion in your area, whatever its level of development!

We are and will remain one species. We will always be dependent upon other forms of life. I like to remind people half-jokingly that, according to the Bible and the book of Genesis, God created the Earth and all of nature first, and then, only on the sixth day, man. This is just one way of remembering that human beings are not the most important part of creation. Even if we vanish off the surface of the Earth or self-destruct, nature will remain. The question is whether or not we, as the human race, want to remain on Earth.

## James Lovelock

# "We can no longer avoid this proof of joint responsibility."

James Lovelock was born in 1919. He is a scientist, researcher, author, and independent environmentalist specializing in atmospheric sciences. He is the inventor of an electron capture detector for use in gas chromatography whose use led to the discovery of widespread pesticide residues throughout the natural environment. He developed the Gaia Theory, which postulates that the earth is a self-regulating system that keeps the climate comfortable for its organisms. The Geological Society of London awarded him the Wollaston Medal in 2006.

Do you remember the torrid heat wave that pummeled the European continent from June to September 2003? It was as if the Sahara had suddenly moved up to the middle of Europe, bringing with it its scorching climate. Since then, many of you have asked whether this was due to global warming. Were these the famous climatic disruptions everyone is talking about? At first, we didn't know anything about it. Perhaps we were simply experiencing a particularly hot summer. But today, climatologists and other experts are practically certain that what we lived through was just a foretaste of a phenomenon that will gradually become more widespread as we move through the century. Global warming is a reality.

Let's try looking at things from the point of view of planet Earth, which the ancient Greeks baptized Gaia, the mother goddess. Let us regard it as a living being, which is what it is, in the same way that you and I are. In what way is the Earth a living being, rather than just something that resembles a huge, inert lump of old rock with a thin layer of air around it and some water on its surface? In fact, it is like those giant trees that grow in California, the sequoias. Their ancient trunks consist of dead wood and it is in their thin layer of bark that all their living material is concentrated. It's the same thing for Gaia. She is capable of self-regulation, of controlling her climate and her chemical composition, so as to provide us with a constant and adequate supply of air to breathe. However, the bad treatment we have meted out to her has simply made her sick.

How have we managed to do enough to damage our planet to put its life in danger? In the course of the last two centuries, which is to say since our society became industrialized, we have changed the face of the Earth in an astonishing fashion. First of all, there's been a demographic explosion. In 1800, there were only a billion of us on Earth, and if we had remained at that figure, we could probably act as we liked today, driving huge cars, using as much air conditioning and heating as we wish, and drinking water till our thirst was quenched without it becoming a real problem. In 2008, however, the world population stands at 6 billion individuals and it will quickly rise to 10 billion. Gaia cannot hold all these people.

The greatest damage that we are causing is the destruction, pure and simple, of our planet's resources. Because we are doing more than just taking what we need to survive; we are destroying them without giving Gaia time to regenerate herself, to "re-make" herself. We have dipped heavily into our capital, as they say. Worse still, as well as considerably weakening the Earth, we are creating the conditions for its death throes, by emitting all sorts of gases (carbon dioxide, methane) that are changing the composition of the air. Although the atmosphere has always contained these gases, their proportions have grown noticeably, to the extent that they now form a sort of blanket above the Earth, a blanket which holds in the heat and makes the climate much warmer than it should be.

If all this had happened when our planet was still young, say 2 billion years ago, it wouldn't have had the same effect. We could have committed these acts with total impunity. But Gaia is growing old! I like to think that we are the same age, and I am nearly ninety. It is estimated that the Earth has only a billion years to live, and as it is already more than 4 billion years old, it now has only a fifth of its life still in front of it. When you are ninety, you are not as resilient as when you were young. We must realize that we are not only behaving badly, we are also abusing an old lady.

Will the ultimate consequence be the disappearance of human beings? No, our species is very adaptable and there will always be people to reproduce, whatever Gaia's general condition. But in years to come, we will have to endure profound climatic upheavals that will have consequences for the survival of the human race. The first, and probably the greatest one, is taking place in the Arctic basin, between Russia and Canada, where ice floats on top of the Arctic Ocean right up to the North Pole. During the summer, this region—which has the same surface area as the United States–is covered with reflective white snow. Almost all the heat that beats down on this layer of ice is reflected back into space and does not bring about any warming of the planet. This floating ice is melting so fast because of the greenhouse effect brought about by gas emissions, and it is estimated that by 2050 it will have disappeared completely. The darkness of the ocean depths will take the place of this reflective whiteness and absorb the extra six months of summer sun.

Once the Earth's temperature has risen by four degrees compared with what is normal in the Tropics today, the huge forests such as those in the Amazon basin (which are already having a rough time) will wither away and become deserts. And thus the planet will lose another great heat sink, for the forests play an important role in regulating terrestrial temperature.

Because of the rising sea level (due to melting ice caps), some countries or parts of the globe, such as Bangladesh and Florida, will be engulfed by the sea. The heat will become unbearable in Southern Europe, which will experience climatic conditions close to those of the Sahara today.

How do we know whether these projections are true, or just the fruit of the over-active imagination of a handful of ecologists? Quite simply because a similar series of events occurred on Earth 55 million years ago. Somewhere in the Arctic region, off the Norwegian coast, there was a huge reservoir of petroleum. A sill (a lava intrusion which never reaches the surface) flowed into this region containing coal and petroleum deposits and caused the latter to evaporate. At least a thousand billion tons of gas were released directly into the atmosphere—almost equivalent to the amount of carbon dioxide emitted by humankind over the past century or so. During this period, which geologists call the Eocene, temperatures rose by about eight degrees Celsius in the northern regions, and five degrees in tropical regions. This warming period lasted more than 200,000 years, and that's a long time—almost as long as the time Homo sapiens have been on the scene. At that time, the Earth knew how to heal itself, of course. These days, it's a very different story, because we have seriously impaired the ability of our planet to recover and fight back. Moreover, the sun's temperature is higher today than it was 55 million years ago.

A combination of different elements will give rise to a much more hostile world than the one we know today. It will become difficult to maintain supplies of fuel, food, and raw materials at the levels which we have come to take for granted, because the products we consume come from all over the world. The smallest nations will find themselves isolated on many levels and will have to make do for themselves. It is at this point that nuclear power will really come into its own, because it alone will be able to ensure a stable and steady supply of the energy we will need. This will enable us to preserve our civilization and prevent it from degenerating and falling under the control of a handful of warlords, for we could easily reach that point. We must not forget that civilizations are not immortal, that they too can disappear.

Several renowned scientists have come up with technological solutions for capturing or diverting 2 to 3 percent of solar radiation. To cite only the most serious ones, these include the positioning of a giant carbon-fiber diaphragm at a given point above the globe, the installation of stratospheric balloons with reflective surfaces, the injection of sulfur particles into the atmosphere by commercial aircraft, and the building of yachts designed to eject seawater spray into the air to form low clouds above the ocean. However, these propositions would only modify the nature of the problem, and in some cases they would be the source of new types of pollution. Imagine a forty-year-old man who learns that he has type 2 diabetes. Medication will help to treat the consequences of his illness without attacking the causes! To be cured, our patient must completely change his lifestyle: stop eating certain foods, eat less, exercise regularly, and lead a quiet life. It's the same for our planet. We must go beyond palliative medication. There's a lot to do, but all is not lost.

One day, we will have to put a definitive end to the use of fossil fuels. Most members of environmental movements say that to avoid the terrible consequences of global warming, we must stop using fossil fuels immediately, economize on energy use, and learn to use renewable resources. This is based on good intentions, but humanity is rather like a drug addict abusing fossil fuels: if he continues, he knows that the drug will end up killing him, but if he stops suddenly, he also risks death. We can't stop using fossil fuels with a single snap of our fingers. The process has to be gradual. It will take some time. As for the idea of completely replacing fossil fuels with renewable energy, this seems absurd to me. These energy sources can only meet a tiny proportion of our needs, with the exception of a couple of regions of the globe. In Iceland, for example, it would be possible to rely solely on renewable energy because that country has a vast reserve of geothermal energy capable of satisfying the needs of its limited population. The same is true of Sweden, to a certain extent, because its hydroelectric network covers half the population's electricity needs. I believe the situation is similar in New Zealand, and surely in other countries, too. We mustn't talk in generalities. Each country must have a variety of energy sources at its disposal that correspond to its needs. But in every region, with the possible exception of Iceland, nuclear power will always be one of the main sources, as well as one of the safest ones at the present time. For contrary to what many people think, there is no doubt that nuclear power is reliable and presents no danger for Gaia.

We must cease intensive and harmful farming practices. The countryside may still look "green," but the type of agriculture we used to practice has given way to agribusiness, to an unnatural industrial agriculture which has exhausted the soil, creating lasting and deep-seated pollution. We must aim for the maximum optimization of crops, which may mean, for example, that we have to stop eating meat, because cattle are raised extremely inefficiently. It takes twenty times the amount of land to raise cattle as it does to obtain an equivalent yield of fruit, vegetables and cereals. Can we allow ourselves this luxury?

When we talk about living things, we mustn't forget that everything that is alive is vulnerable to some kind of pathology. As we have seen, this concept applies to Gaia's imbalances, but it is also valid for another superorganism that we have created: the city. Between the Middle Ages and the first industrial revolution, in around 1800, although life in the city wasn't particularly healthy (conditions were insalubrious and there were a great many epidemics) the city itself was a healthy organism because it was efficient and not very spread out. The countryside was very close, and wherever you were in town, you could easily reach the country by walking. Comparisons with the modern megalopolises speak for themselves: 95 percent of the world's population live in an urban environment, in cities so dense and sprawling that we have lost all contact with nature. Rare are the people who have the opportunity to leave the city to walk in the country, recharge their batteries and regain the sense of wonderment that is indispensable if we are to respect the environment. Our thoughts are invaded with roiling urban images, and our vision shrinks to such an extent that it eventually restricts itself to the world of the city. Our bonds with Gaia are stretched to the limit. Even so, we must never forget that we are part of her and that we must treat her with respect.

Can my action as an individual really make a difference on a global level? Yes. In the past, the universe evolved thanks to individual organisms which successfully played their different roles. We would not have the air we breathe today if certain microorganisms hadn't changed on an individual level a long time ago, and if that change hadn't been beneficial to their development. This applies to all individual action: if it succeeds and brings some kind of benefit, it will be copied, will multiply and have a considerable effect, an effect on a global level. When people speak of pollution, they generally accuse petroleum companies, governments, etc. They aren't completely wrong, but responsibility also rests with individuals, who have become consumers above all else. We cannot avoid this evidence of responsibility.

It is often thought that science is neutral from an ethical point of view and that it has no opinion about what is right or wrong. Mathematics has demonstrated that to produce stable models, it is necessary to introduce major physical constraints, whatever the system being studied and its size. In other words, stability depends upon the existence of these constraints. I can't help but make an analogy with the functioning of a family. If the children in a family are not surrounded by constraints (reasonable ones, it goes without saying), the life of that family is condemned to chaos. This is a fundamental law that applies to our planet's ecosystems as well. By dint of research and experience, I have formed a scientific theory that possesses a certain moral dimension, speaks of punishment, of good and evil. This is the Gaia hypothesis, which can be defined as a new theory of evolution. It does not contradict the older, Darwinian one, but completes it by putting Gaia at the center of the system. It associates the evolution of organisms with that of their natural environment—the air, the oceans and the rocks—in such a way as to form a system which evolves as a single entity.

Contemporary science is moral in the sense that it affirms that we are indebted to the Earth. Gaia has rules by which we must abide. Every species that degrades its environment, which inflicts damage upon it of whatever kind, renders this environment more hostile for its descendants and will not be selected for survival. If this species does not change its behavior, it is doomed to extinction. This resembles a form of punishment in many respects. On the other hand, any species that cares for its environment and works to make it better for its descendants will triumph and inherit the Earth.

The clock is ticking. We must accept our responsibilities and make the right choices. The prospects may seem limited, but if we act reasonably, humanity still has a future.

## Edgar Morin

# "Today, least must mean best."

To tackle the complex question of our relationship with nature, we must first establish some essential benchmarks, and then patiently take up the Ariadne's thread of the past.

The science of ecology was born in the nineteenth century. Its root, the Greek word *oikos* means "house," a house of life built from relationships between plants, animals, minerals, meteorological and geographical conditions, etc. It was shortly before World War II that the concept of "ecosystem" emerged in this science. What is an ecosystem? It is the spontaneous organization of microbes, animals, minerals, and plants in certain geographical, geological, and meteorological conditions. The most interesting aspect of it is that even though there is no brain or any other organ of command or control, these complementary and antagonistic sets of organisms create an organization endowed with a certain self-regulatory ability. The more varied the living species, the more complex the ecosystem and the greater its future development is likely to be.

Four billion years ago, when there were only single-celled organisms on Earth, one could not yet speak of ecosystems. It took the proliferation of plants and animals for ecosystems to appear. It is the combination of all these constantly interacting ecosystems that forms what we call the biosphere.

The organization of ecosystems requires solidarity in both life and death. This is what ecologists call the trophic network, or food chain. Insects feed on plants and birds eat seeds, herbivores, such as cows, graze on grass, and carnivores devour other animals. When they die, all of them, even the mineral salts, feed the chain via a process of decomposition. The richer this cycle is, the richer the ecosystems and biosphere. The advent of the great apes, and finally the hominids, our ancestors, required extremely favorable ecosystems and a tropical context. However, a huge link then emerged within this chain: human intervention and its disastrous consequences! It was only in the late 1960s and the 1970s that we started to realize that there was an environmental problem which did not concern individual ecosystems but was far more global, affecting the whole of the biosphere.

It was during this period that the academic Paul Ralph Ehrlich announced the death of the ocean. A death that would not only be a physical and biological problem due to the discharge of waste, pollutants and toxic substances, but which would also have incalculable human consequences, since fish and marine mammals would disappear, along with all the other benefits we derive from the ocean.

At the same time, alarm bells were also rung by the Meadows report, commissioned by the Club of Rome, which showed that the technical, economic, urban and industrial processes of our civilization were bringing about a general degradation of the biosphere and, in consequence, were having a backlash effect on humanity.

These warnings went largely unheeded. However, since then we have witnessed some disturbing phenomena and a whole series of catastrophic events, including acid rain, Seveso, Chernobyl, the degradation of the ozone layer, global warming and the melting of the Arctic and Antarctic icecaps. Certainly, the Earth is capable of resisting, as James Lovelock has passionately demonstrated with his Gaia hypothesis. But just how resilient is she? We simply don't know.

Over the decades, our bond with the biosphere has become clearer to us. For a long time, we lived with the notion that, thanks to its technical civilization, humanity was not only capable of conquering and dominating nature, but had a mission to do so. It saw living things only as things to manipulate. This separation between nature and humanity in the western world is strongly rooted in our religious history. According to the Bible, man was made by God in His own image and therefore carries a spark of divinity within him which distinguishes him from the animal world. The Gospels tell us that the son of God, Jesus, was made flesh to bring his message of compassion only to humans. In Buddhism, however, empathy with others is not tied only to human suffering, but encompasses the animal, plant and mineral kingdoms.

In the seventeenth century, imbued with this supposed superiority, René Descartes declared that the goal of science was to make man the "master and possessor of Nature." Subsequently taken up by the Comte de Buffon and later Karl Marx, this concept formed the basis of the illusion of our western civilization until the 1980s. This presumptuousness is driving us toward suicide, for the more we possess nature, the more we are possessed by those unbridled forces that threaten our destruction. This determination to conquer the world became definitively null and void when, as recently as the 1960s and 1970s, we discovered that our solar system was just one tiny system in a peripheral galaxy of a gigantic universe. And in any case what does it mean to be the master of nature when we are the ones responsible for its degradation? Four billion years of evolution have produced the extremely rich, complex, and diverse reality that is our biosphere, life on Earth. Anything that undermines this wealth indirectly undermines us. This is the thinking of all those who work to preserve threatened animal species, even when these species are regarded as predators of humankind. They sense that diversity is one of the treasures of life, of humanity: just as the diversity of cultures and of thought is one of our treasures. However, by dint of responding to short-term emergencies, we have tended to lose sight of the bigger picture. Only now have we come to realize the need to reflect upon the relationship between humanity and the living planet.

This relationship has considerably evolved over the course of human history. For tens of thousands of years, humans formed small, seminomadic societies, which lived of the fruits of their hunting and gathering without changing the equilibrium of the natural environment.

From the eighth millennium B.C. onwards, large, "historical" agricultural societies started to emerge. Taking the form of a city-state or empire, they gave rise to the development of cities, armies, and slavery. It wasn't until this period that farming began to modify the relationship between humanity and nature, for cities created a new environment, outside of nature. This is when the damage started, having an impact not only on the environment, but also on the civilizations themselves, given that the history of humanity is littered with tales of the massacre and destruction of these cultures. This was a new factor, since previously all the great traditional civilizations had developed belief systems in which the role of humans was not only to be at one with nature, but also to help it. The Aztecs, for example, made regular sacrifices of young girls and boys, not out of cruelty, but so that their energy would allow the sun to regenerate itself. Humanity had to help nature, and in exchange nature rewarded humanity.

City life and farming went hand in glove. It was the development of cities that made it necessary for the land to provide food in ever greater abundance. The city furthered the development of farming methods, but equally provoked numerous conflicts. One people would attack another to steal its agricultural resources in time of drought, flood, epidemics, et cetera. Wherever populations were concentrated, a genuine social metamorphosis occurred. Then the great empires of antiquity were destroyed, including the Roman Empire. The profound relationship which bound humankind and nature together continued to disintegrate little by little. We had to wait until the end of the eighteenth century for the first stirrings of consciousness. The message of Jean-Jacques Rousseau, that great apostle of sensibility, was that human beings should return to their true nature and to a harmonious life within the bosom of nature! The Romantic movement exalted this reunion in its poetry, music, and painting, but would prove to be only a sideshow alongside the formidable urban, technical, and industrial development that took place throughout the nineteenth century.

The real, radical transformation occurred in Europe with the two industrial revolutions of the nineteenth and twentieth centuries, which developed technology and capitalism further. The urban world became more dense, more extensive, and humanity continued to loosen its ties with nature all the more rapidly due to rural depopulation. Farm laborers wanting to escape from the constraints and poverty of rural life flocked to the city in droves. This phenomenon was hastened by the mechanization of agriculture, linked to the development of large farms and the destruction of smallholdings. Urban life, already prosperous, grew even more so with advances in hygiene and medicine, and the elimination of the major epidemics. Demographic growth went unchecked, which is how we come to find ourselves today living in megalopolises of 8 to 10 million inhabitants, where nature is nothing but an abstraction! Our age is also still marked by uninterrupted wars. The great empires of the modern world (Austro-Hungarian, Ottoman, Soviet) collapsed one after the other in the twentieth century in the wake of increasingly murderous conflicts, culminating in the mass slaughter that defined the two World Wars.

Is this history humanity's inevitable destiny? In order to escape it, we must move beyond industrial empires and nation states, and construct a global society that will give rise to a humanity without war, fully aware of its terrestrial destiny.

We are on a path which will lead us to catastrophe. This path is called growth. Even if it is moderated by the notion of sustainable development, that is, taking account of the environmental issue, we will still expect technology and the economy to bring us the very best. The truth is that we are developed materially, but we are psychologically and morally immature. Moreover, in many of what we used to call the under-developed countries, we can see just how far our technical and economic development, held up as a model, has fomented corruption and destroyed communities and traditional bonds. We have reached a point where we must not only reconsider our relationship with nature, but also reconsider our relationship with ourselves, with our own civilization. Our individualism and the hyperspecialization of our societies (the two great triumphs of modernism) have become a very real danger, preventing us from achieving common good. For we are now forced to face a common destiny, because we are all threatened by the death of the biosphere and the proliferation of weapons of mass destruction, chief among them the nuclear bomb.

In the end, each of us is possessed by what we possess. We must stop consuming for consumption's sake. Quantity has taken the place of quality. Today, least must mean best. We must face up to this paradox and change. Unbridled growth makes no sense in a limited world. It is irresponsible to focus solely on the depletion of mineral resources. This is to forget that their gradually increasing costs will create a whole series of economic and social disturbances. We should remember, for instance, that the huge economic crisis of 1929 resulted in generalized unemployment worldwide—not least in Germany, already humiliated by its defeat in 1918—and brought Hitler to power in a totally legal fashion. The fascist and totalitarian regimes of the twentieth century were born out of the convulsions that were the First and Second World Wars.

The growing dissatisfaction within our civilization, the absence of prospects and the tensions due to the scarcity of sources of energy and drinking water may also spark movements calling for a return to narrow nationalism and closed societies. All the current conflicts are to do with identity, religious affiliation, and nationalism and only serve to worsen the disaster, especially given the threat of nuclear weapons. It is an explosive situation. New forms of religious fanaticism may develop, old ones be revived.

Edgar Morin is a French sociologist and philosopher. He was born Edgar Nahoum to Sephardic Jewish immigrants in 1912. When Germany invaded France in 1940, he became an early member of the Resistance in Toulouse, took the wartime pseudonym Morin, and wrote *The Year Zero for Germany* in 1946 for *La Libération*, which reported on living conditions in that country. He joined the National Center for Scientific Research in 1950 and is currently its director emeritus. An institute at the School of Advanced Studies in the Social Sciences in Paris bears his name. He is renowned for his interdisciplinary approach to the social sciences, and has received honorary doctoral degrees from sixteen universities around the world. He is the author of over sixty books, including *Homeland Earth: A Manifesto for the New Millennium*, 1999, *and On Complexity*, 2008.

No one has a crystal ball, but we will inevitably experience further convulsions, regressions, and collapses of democracies. We are heading toward a bottleneck that will inevitably lead to a very violent conflict, in ten or twenty years' time.

There is an interdependent relationship between environmental, economic, social, and psychological phenomena. We must break with the ideology of growth as the solution, since growth has now become our main problem. We must now find and explore other paths.

We need to reconsider the unity and diversity of the human species. This vital process of reflection requires educational reform. We must, for example, cease to teach subjects in a fragmentary manner; we must present an overall picture, showing the full complexity and all the interconnections. Some people are already working on this approach, but we don't yet have institutions where this kind of thought can be developed. In 1810, the modern university was created in Berlin because the great philosopher of language Wilhelm von Humboldt had developed a doctrine which broke with the theological thinking of the Middle Ages. This modern university, which trains compartmentalized minds, is universal today. It is this university which must now be reformed.

We must relearn how to live. This need is emerging, unconsciously, within our civilization. For it was in the Western world (in the United States and Europe) that the need first emerged for a new relationship between body and soul. We began drawing upon yoga, Zen Buddhism, and other Eastern practices to create balance within ourselves. At the same time, many people began rediscovering the countryside, trying to satisfy their need for natural spaces with more frequent weekend vacations in the country, where a more rustic life can be experienced. Urban existence obeys the logic of the artificial machine: it is timed, implacable, and deterministic. It is no longer subjected to a living and human logic, but to a pace which is psychically tiring and impoverishing. The problem of the city is two-sided, because great civilizations are inseparable from megalopolises. We can no longer set the city against the countryside. New sorts of development and a new relationship must be created between them, not only by rehumanizing the city but also by repopulating rural areas and developing a new way of life there free from its former constraints. Telecommuting and the Internet will surely allow for decentralization and movement away from the cities.

Since the 1970s, we have been witnessing a return of people to rural areas. We are trying, by various means, to escape urban asphyxia. Deserted villages only need a small population to support the reopening of a bakery or a café for young retirees, or even young people, who wish to return to the countryside to live. Organic farming, which involves small or medium-sized operations, forms an important part of this venture. These achievements are, however, few and far between: the change must be more radical. For the time being, these models of alternative lifestyles are marginal or provisional. In human history, new ventures have always started out as deviations from the norm that they later develop in turn. This was the case for all the great currents of thought: Jesus, Buddha, Mohammed, Karl Marx. Their ideas spread and the deviations become trends, which is to say forces that act within and transform society. Humanity's major problems can be resolved by other means than growth, calculation, the economy, and industrial mechanisms. We can develop a civilization that allows people to live better, in relationships based on understanding, friendship, and sharing.

We can't predict the nature of this metamorphosis, but that, I believe, is the principle behind every crisis. A crisis occurs when uncontrolled processes develop and can no longer be regulated. When one can no longer foresee what is coming. This is the progression of uncertainty. Only three paths can make it avoidable.

The first is the restoration of the status quo, which is unimaginable now, since we are trapped in an unending dynamic process. The second is regression, which consists in abandoning the richest and most complex forms of political, social, and democratic life, and returning to former states: dictatorships, totalitarianism, or even the "remedievalization" of societies. The last hypothesis is progression: the crisis reveals that the self-evident facts on which the functioning of an efficient system is based are no longer valid. We must therefore change our ideas and invent. Crises stimulate our powers of imagination and invention.

At the end of the eighteenth century, the German poet Friedrich Hölderlin wrote "where danger is, grows the saving power also." Why? The greater the danger, the greater our awareness of it, and the better and more original the possibilities for salvation.

Personally, I am not discouraged. There is a horizon to despair, beyond which lies hope. Hope can prompt us to protest the mechanisms of the current world and imagine another form of existence. We have reached an extremely critical period, wherein humanity is no longer managing to be born as humanity. And in this agonizing effort to be born, humanity risks dying. This, I believe, is the moving and exciting aspect of our age.

In the past, the planet has gone through periods of great crisis which have all led to new developments. Only the future will tell us whether we have achieved awareness and moved to act it in a way that will save us.

*To Héléna and Gaspard*

Acknowledgments

First, to Héléna, who bore this project along with me and supported my absences, to my parents and my sister, unflagging supporters from the very beginning. Thanks also to Alexandre, Renaud, and Arnaud of gallery Acte2; to my editor Pierre-Henri; to Florence Faisan and Juliet Piper for their insightful advice; to Philippe Chaume, his team, Pierre Barougier and Thiphaine Bireauteau—I have such fond memories of the important time we spent together; to Nicolas Grandmaison for his unsurpassable good humor, his permanent investment in this book and his imitations of Francis Cabrel—thanks Mec; to the production teams in the Ukraine, United States, and Brazil; to the companies that opened their doors to us; thank you also to Jean-Sébastien Stehli for his ear; to Sandrine Berthaud for her competence and availability; to Emmanuel Lahaye, the youngest and most polite designer in the galaxy; and thanks also to Didier and Ericka of photographie.com; and to all those who made me want to pick up a camera. Also a very special thanks to Olivier Bourgeois for his enthusiasm, energy, and pugnacity, without which this project could never have come together. I hope to have earned the faith you have shown in me. Through these few words, I would like to express all my gratitude and my friendship.



This work was originally published in different form in a bilingual edition as *Here to Stay/Nous Resterons sur Terre* by Verlhac Editions, Paris, in 2008.



Delsaux, Cédric.
A common destiny : a photographic journey through a changing world / Cédric Delsaux ; introduction by Bill McKibben.
p. cm.
ISBN 978-1-58093-255-4 (hardcover)
1. Landscape photography. 2. Industrial sites—Pictorial works. 3. Environmental degradation—Pictorial works. 4. Nature—Effect of human beings on—Pictorial works. 5. Environmental conditions—Pictorial works. I. Title.
TR660.5.D444 2009
779.092–dc22
2009018839

Printed in China

10 9 8 7 6 5 4 3 2 1

First edition

Design by David Blankenship and Emmanuel Lahaye

www.monacellipress.com